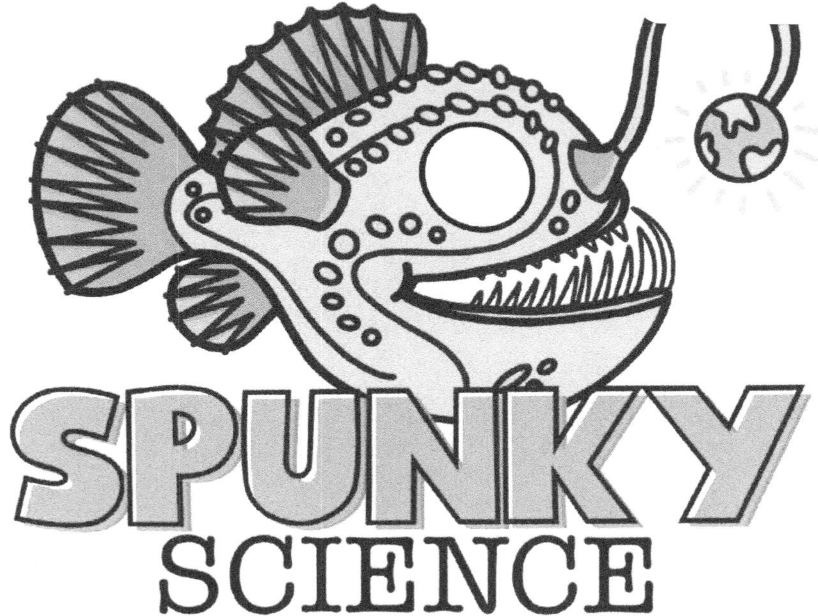

SPUNKY SCIENCE

CAN:

- ✓ Make copies for your students for educational use
- ✓ Print content in different forms such as a booklet
- ✓ Print in various sizes to fit your needs
- ✓ Post content on a school-based platform for student use or reference

CAN'T:

- ✗ Distribute digital or copies to others without an additional purchase
- ✗ Remove Spunky Science logo or copyright
- ✗ Resell or redistribute in any way other than originally intended by Spunky Science

4 TYPES OF TISSUES

IN THE HUMAN BODY

CONNECTIVE

Tissue that connects, supports, binds, or separates other tissues or organs.

EPITHELIAL

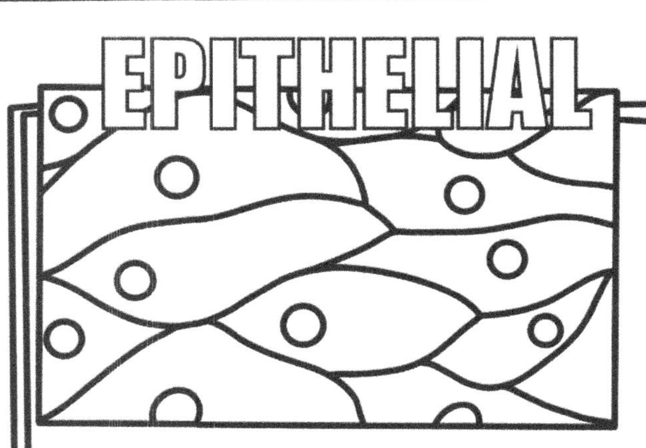

Tissue that forms the covering on all internal and external organs. They serve as a protective barrier as well as secreting and absorbing substances.

MUSCLE

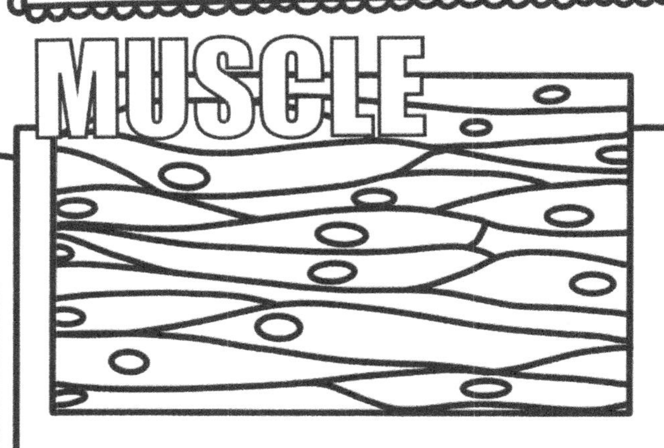

Tissue that attach to bones or internal organs and blood vessels and are responsible for movement.

NERVOUS

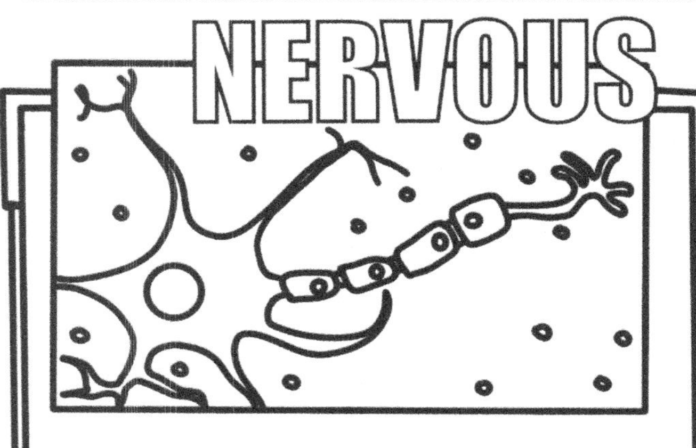

Tissues that are the main component of the nervous system-made of primarily neurons and glial cells.

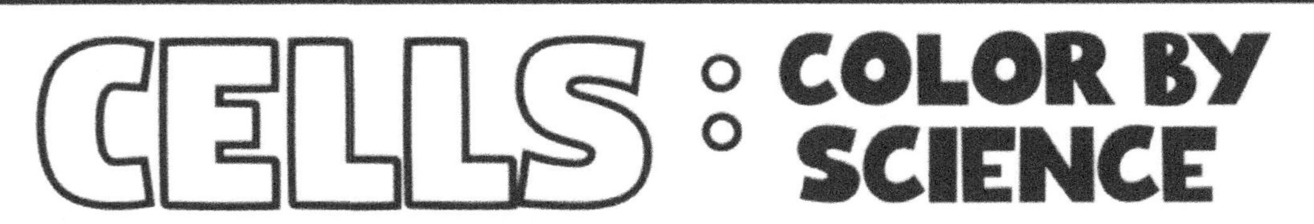

Test your knowledge of cells, by coloring each space the appropriate color to show if it is referring to a Prokaryote or a Eukaryote.

PROKARYOTE	EUKARYOTE
BLUE Shades	PINK Shades

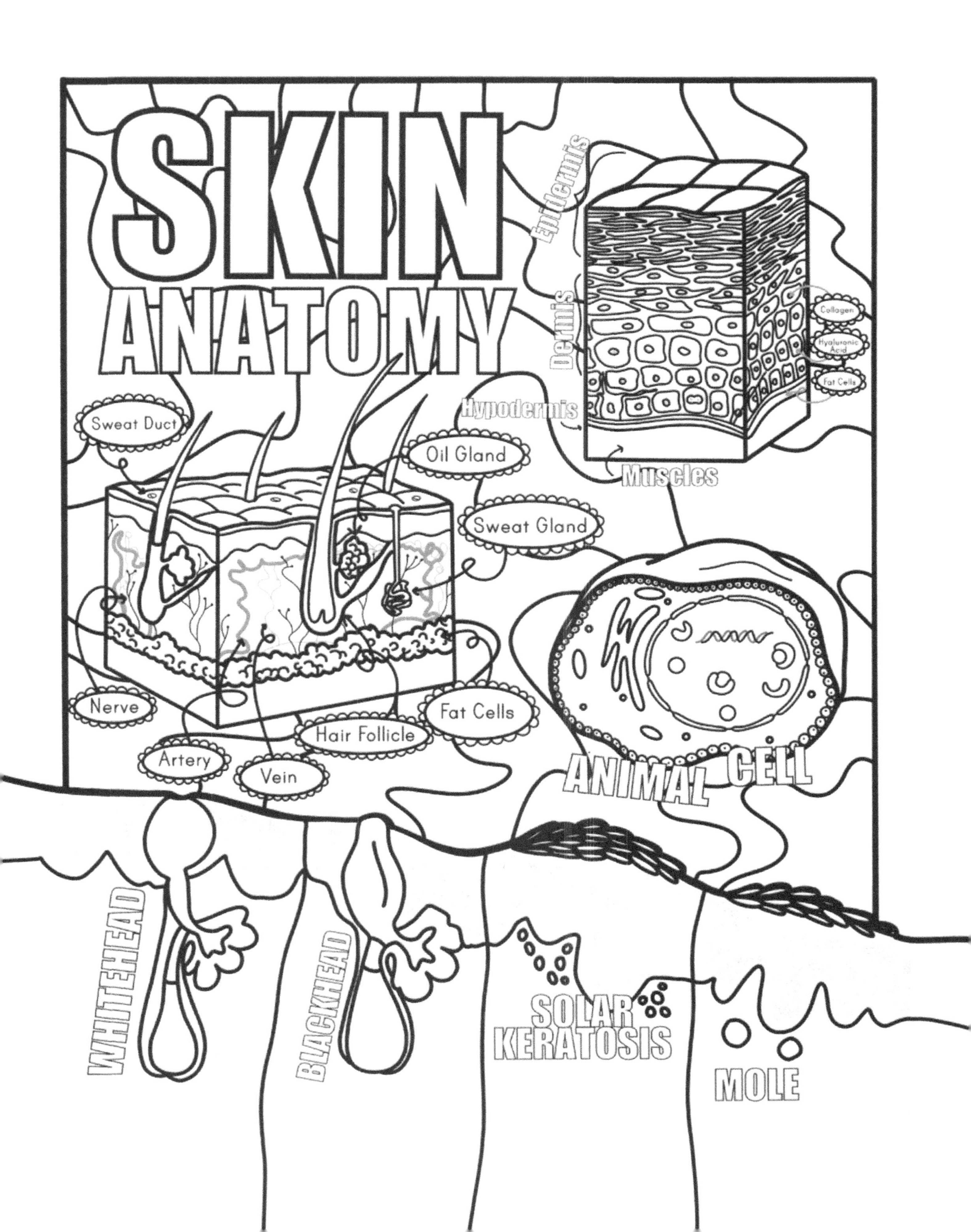
SKIN
ANATOMY
Epidermis
Dermis
Hypodermis
Muscles
Collagen
Hyaluronic Acid
Fat Cells
Sweat Duct
Oil Gland
Sweat Gland
Nerve
Fat Cells
Hair Follicle
Artery
Vein
ANIMAL CELL
WHITEHEAD
BLACKHEAD
SOLAR KERATOSIS
MOLE

Levels of biological organization

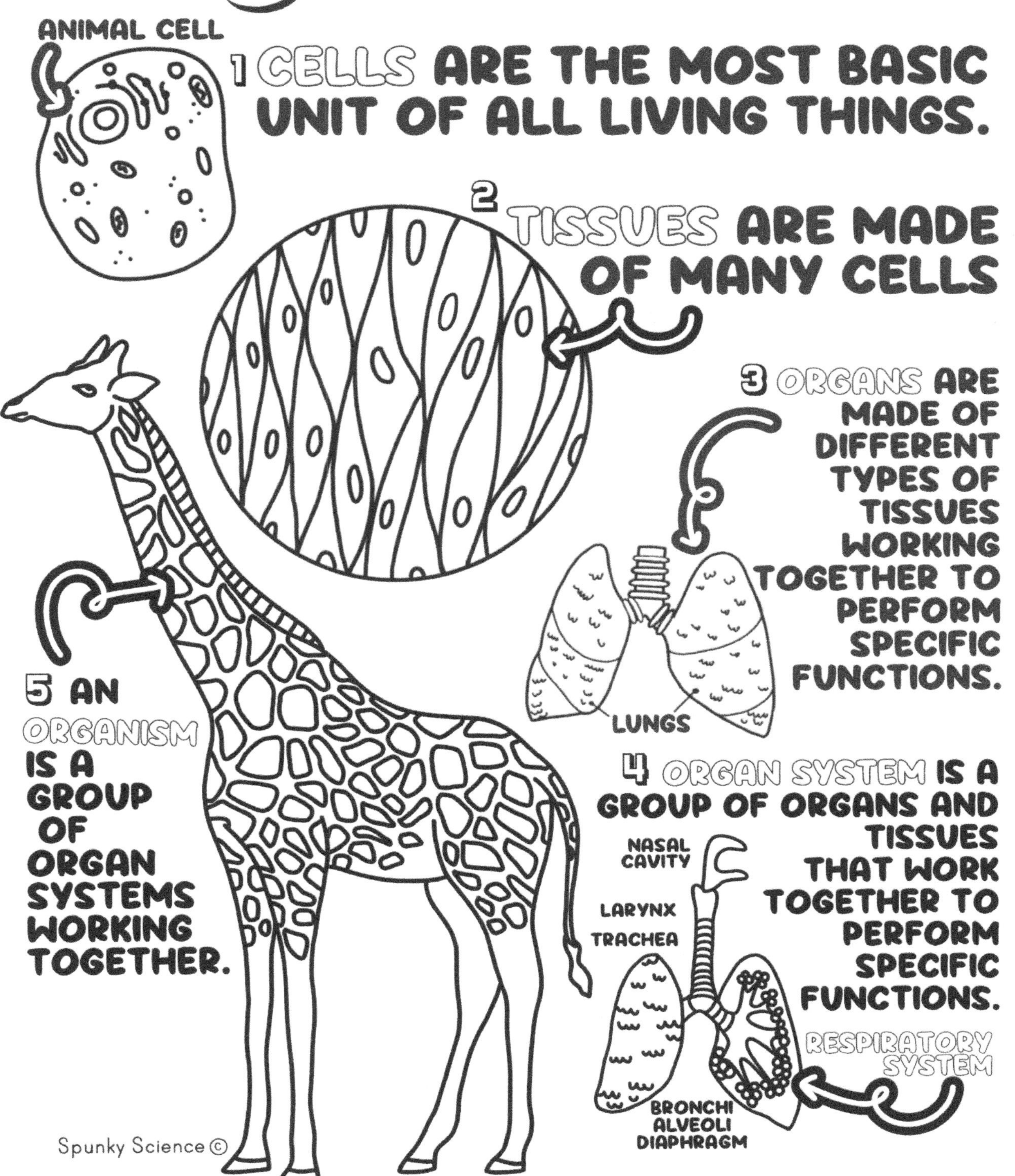

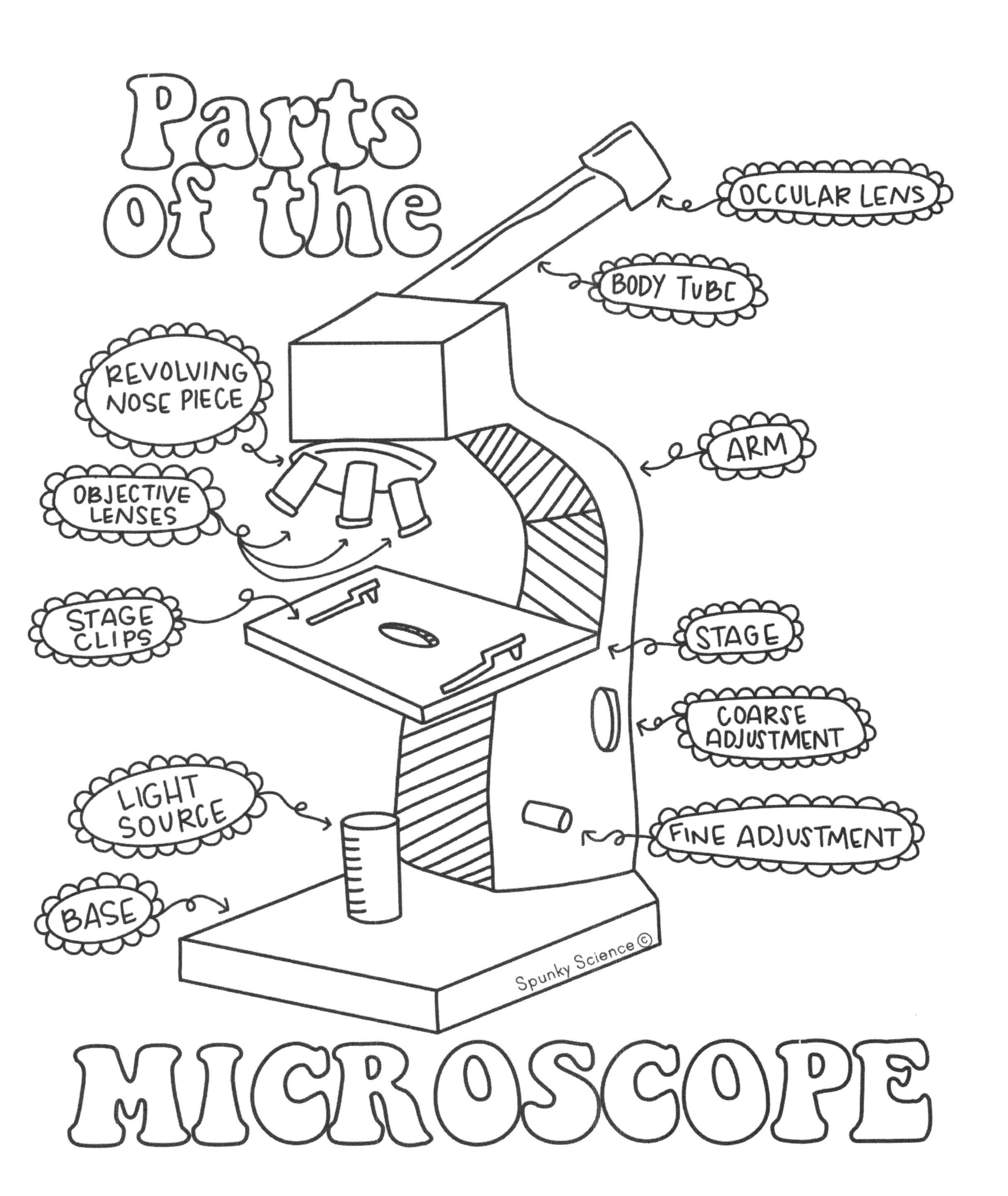
Parts of the
OCCULAR LENS
BODY TUBE
REVOLVING NOSE PIECE
ARM
OBJECTIVE LENSES
STAGE CLIPS
STAGE
COARSE ADJUSTMENT
LIGHT SOURCE
FINE ADJUSTMENT
BASE
Spunky Science ©
MICROSCOPE

CELL THEORY

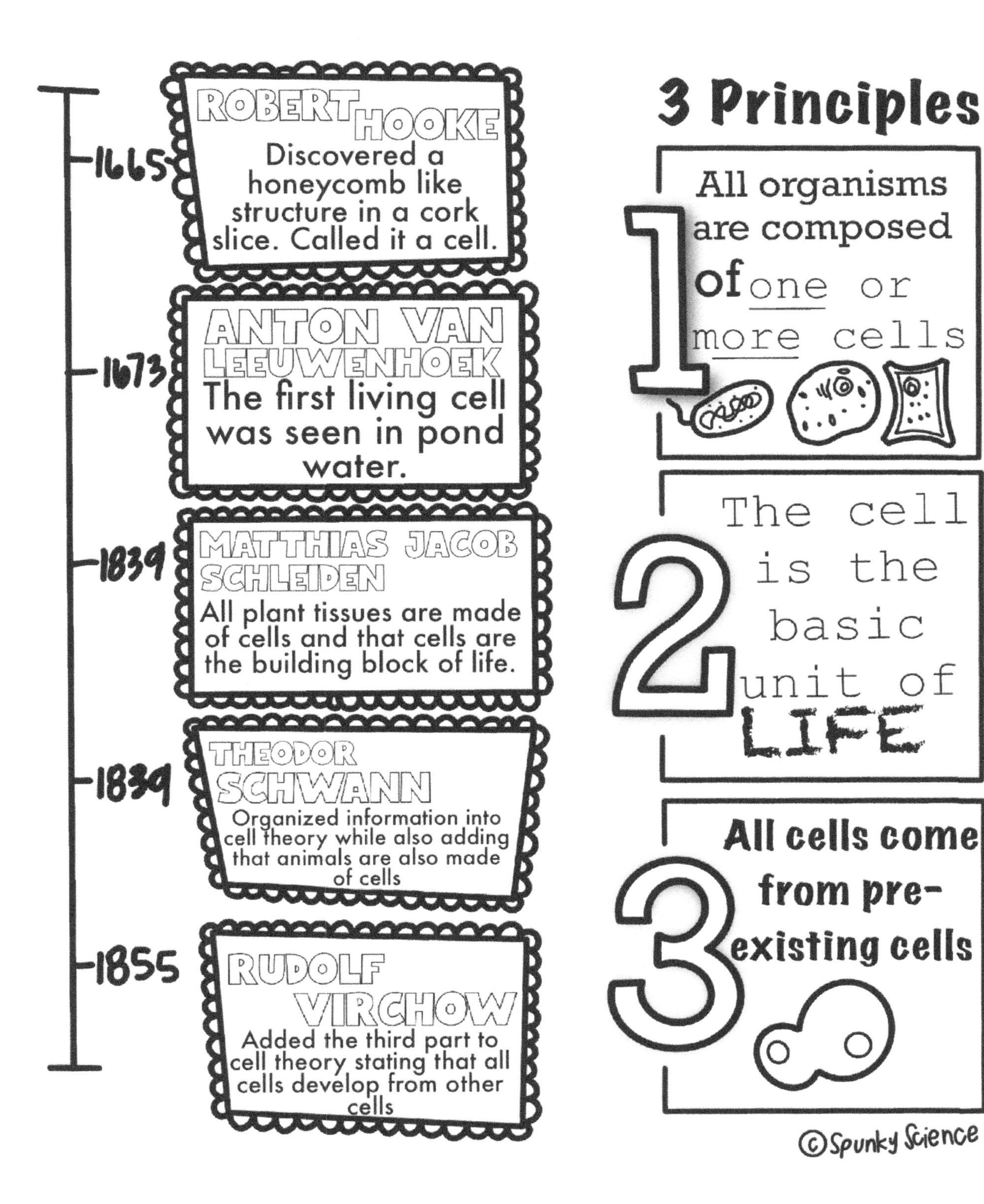

1665
ROBERT HOOKE
Discovered a honeycomb like structure in a cork slice. Called it a cell.
1673
ANTON VAN LEEUWENHOEK
The first living cell was seen in pond water.
1839
MATTHIAS JACOB SCHLEIDEN
All plant tissues are made of cells and that cells are the building block of life.
1839
THEODOR SCHWANN
Organized information into cell theory while also adding that animals are also made of cells
1855
RUDOLF VIRCHOW
Added the third part to cell theory stating that all cells develop from other cells
3 Principles
1
All organisms are composed of one or more cells
2
The cell is the basic unit of LIFE
3
All cells come from pre-existing cells

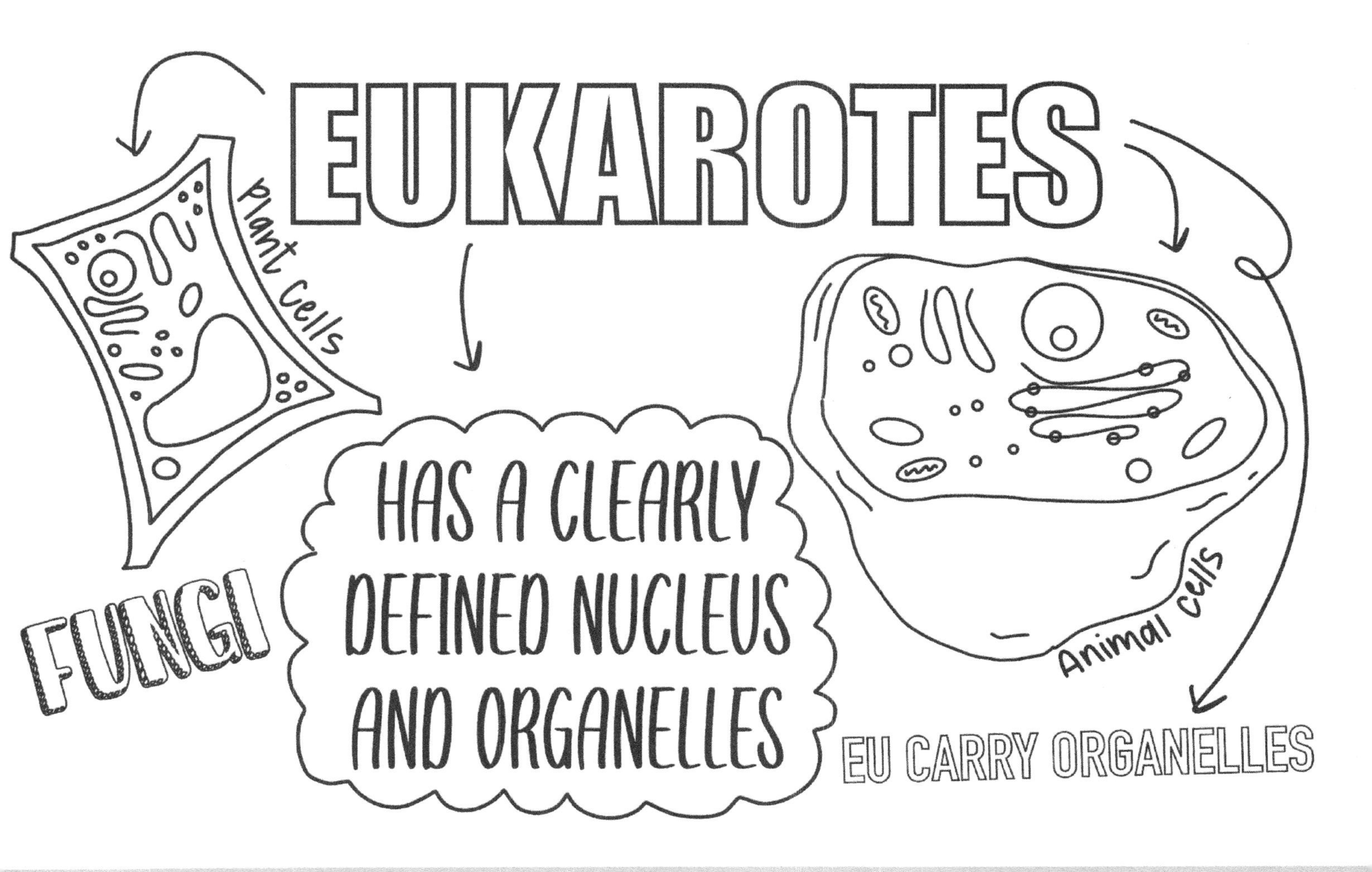
EUKAROTES
Plant Cells
FUNGI
HAS A CLEARLY DEFINED NUCLEUS AND ORGANELLES
Animal Cells
EU CARRY ORGANELLES

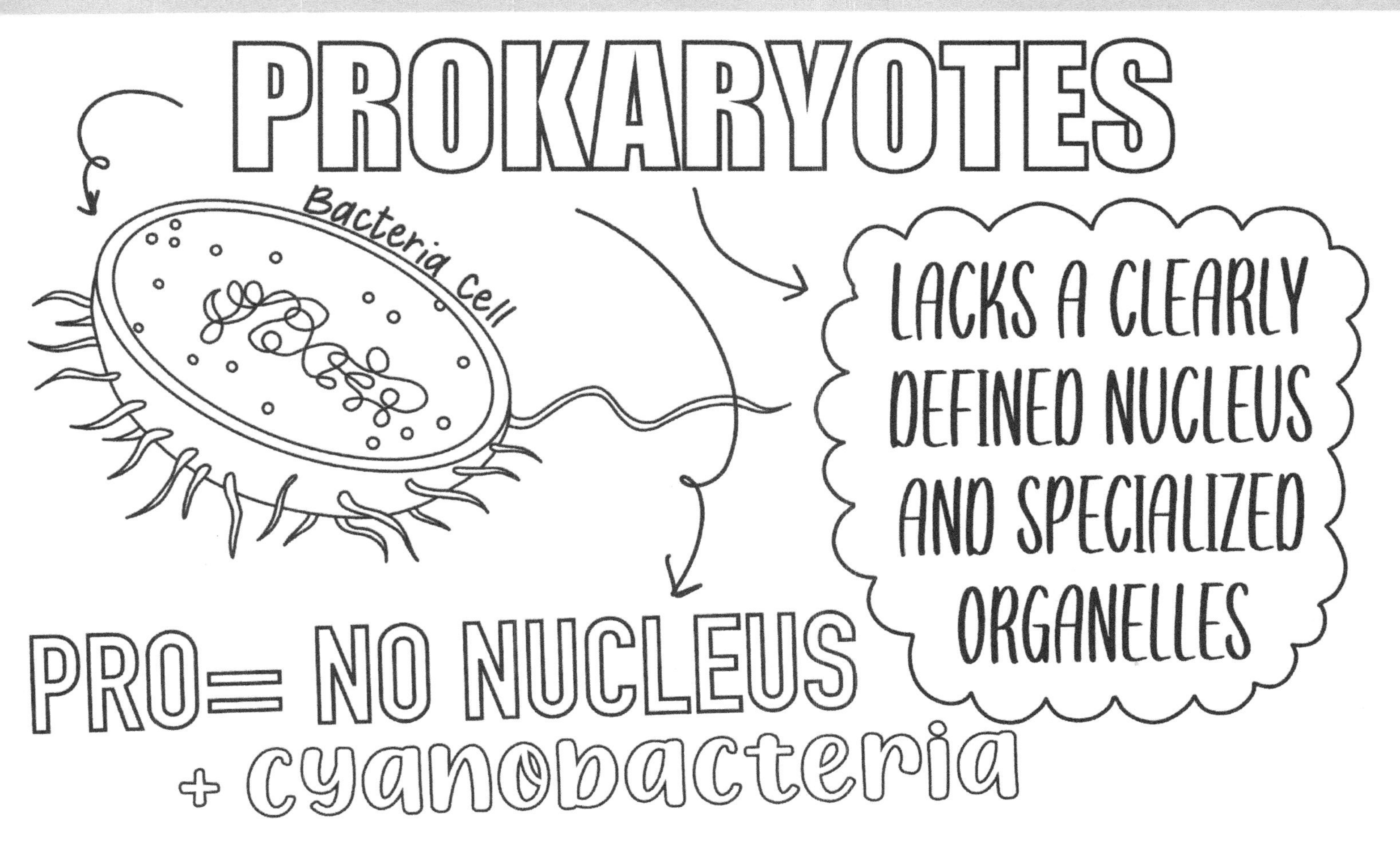
PROKARYOTES
Bacteria cell
LACKS A CLEARLY DEFINED NUCLEUS AND SPECIALIZED ORGANELLES
PRO= NO NUCLEUS
+ cyanobacteria

BACTERIA CELL

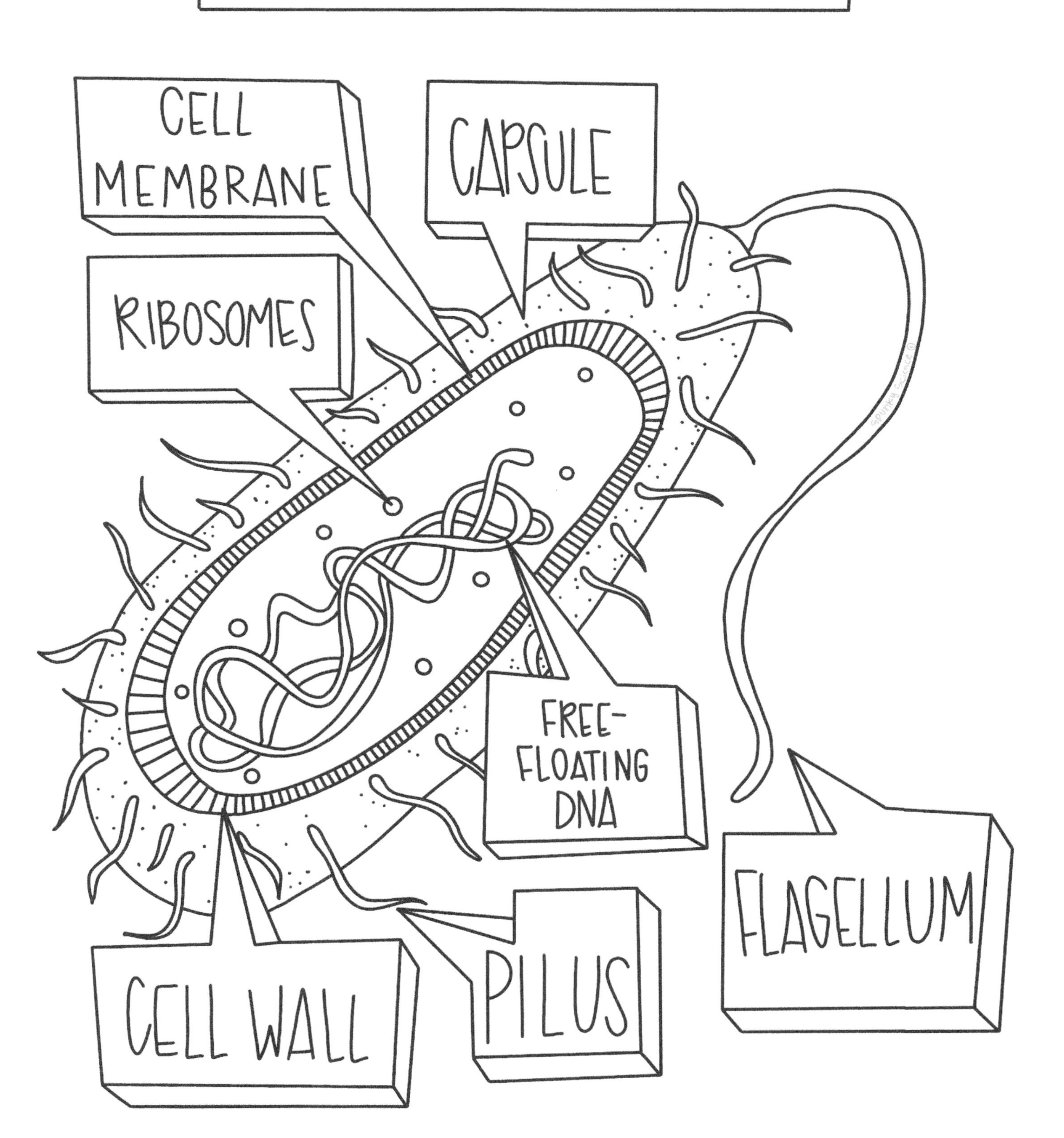
CELL MEMBRANE
CAPSULE
RIBOSOMES
FREE-FLOATING DNA
FLAGELLUM
CELL WALL
PILUS

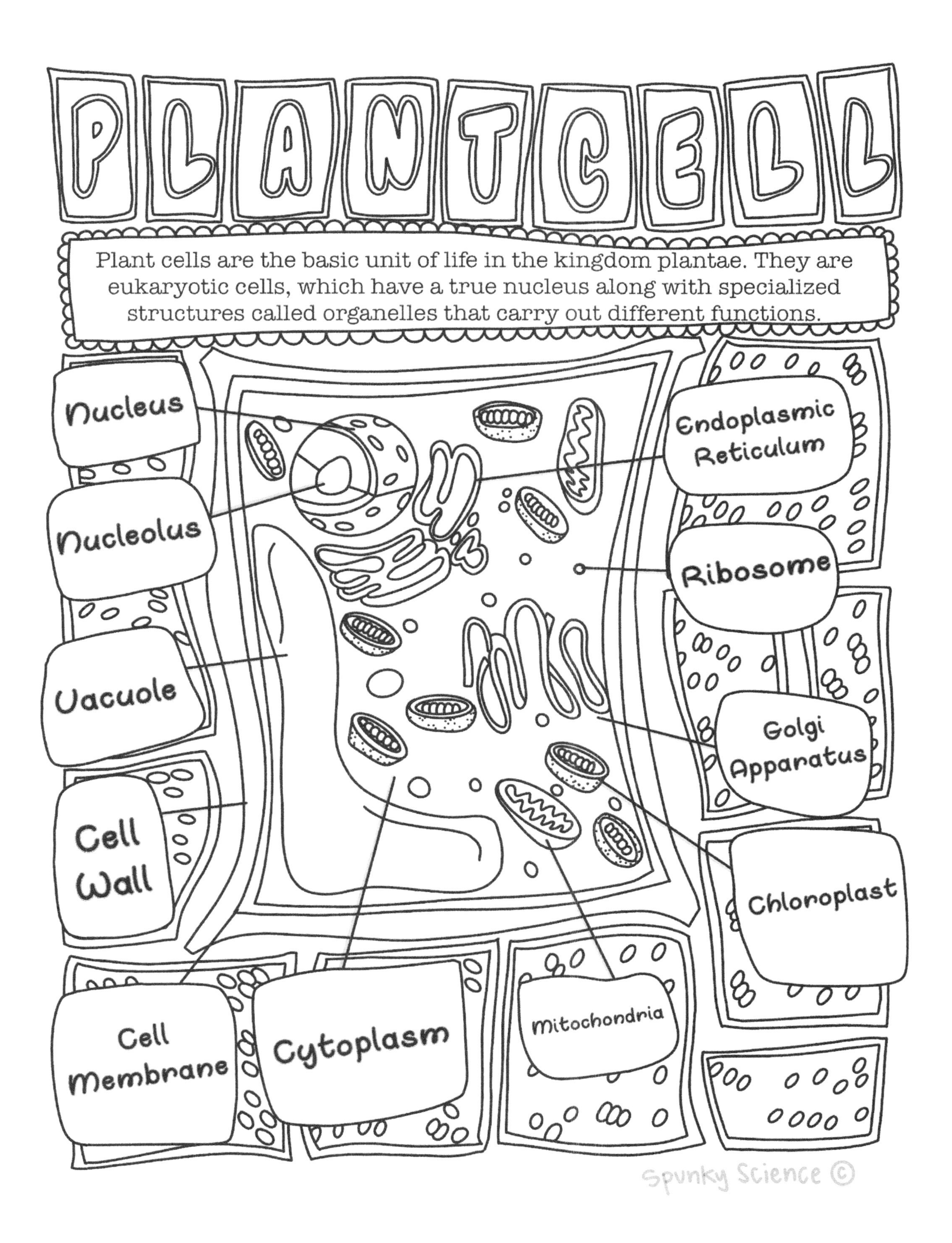
PLANT CELL
Plant cells are the basic unit of life in the kingdom plantae. They are eukaryotic cells, which have a true nucleus along with specialized structures called organelles that carry out different functions.
Nucleus
Nucleolus
Vacuole
Cell Wall
Cell Membrane
Cytoplasm
Mitochondria
Endoplasmic Reticulum
Ribosome
Golgi Apparatus
Chloroplast

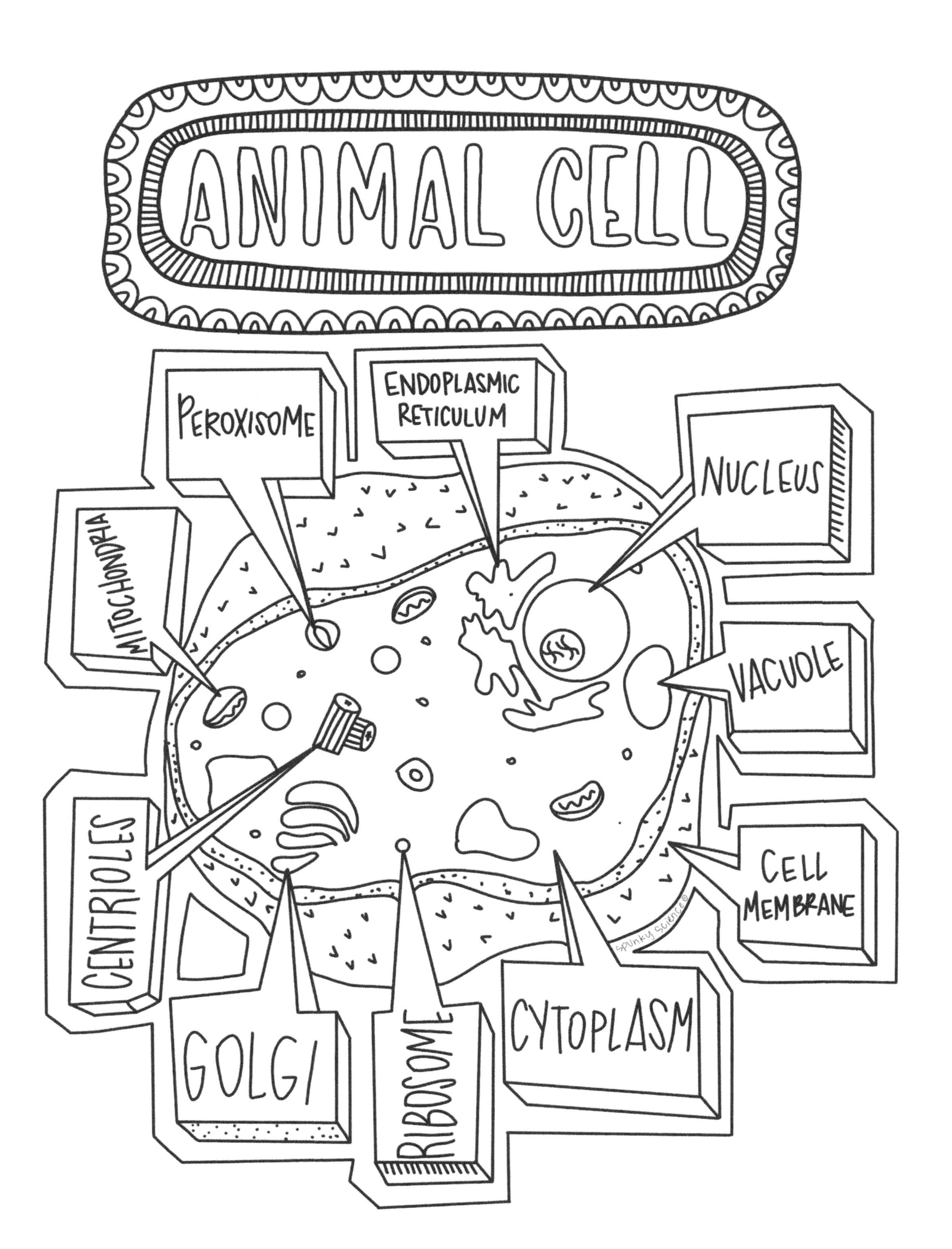

ANIMAL CELL
PEROXISOME
ENDOPLASMIC RETICULUM
NUCLEUS
MITOCHONDRIA
VACUOLE
CENTRIOLES
CELL MEMBRANE
GOLGI
RIBOSOME
CYTOPLASM
spunky science©

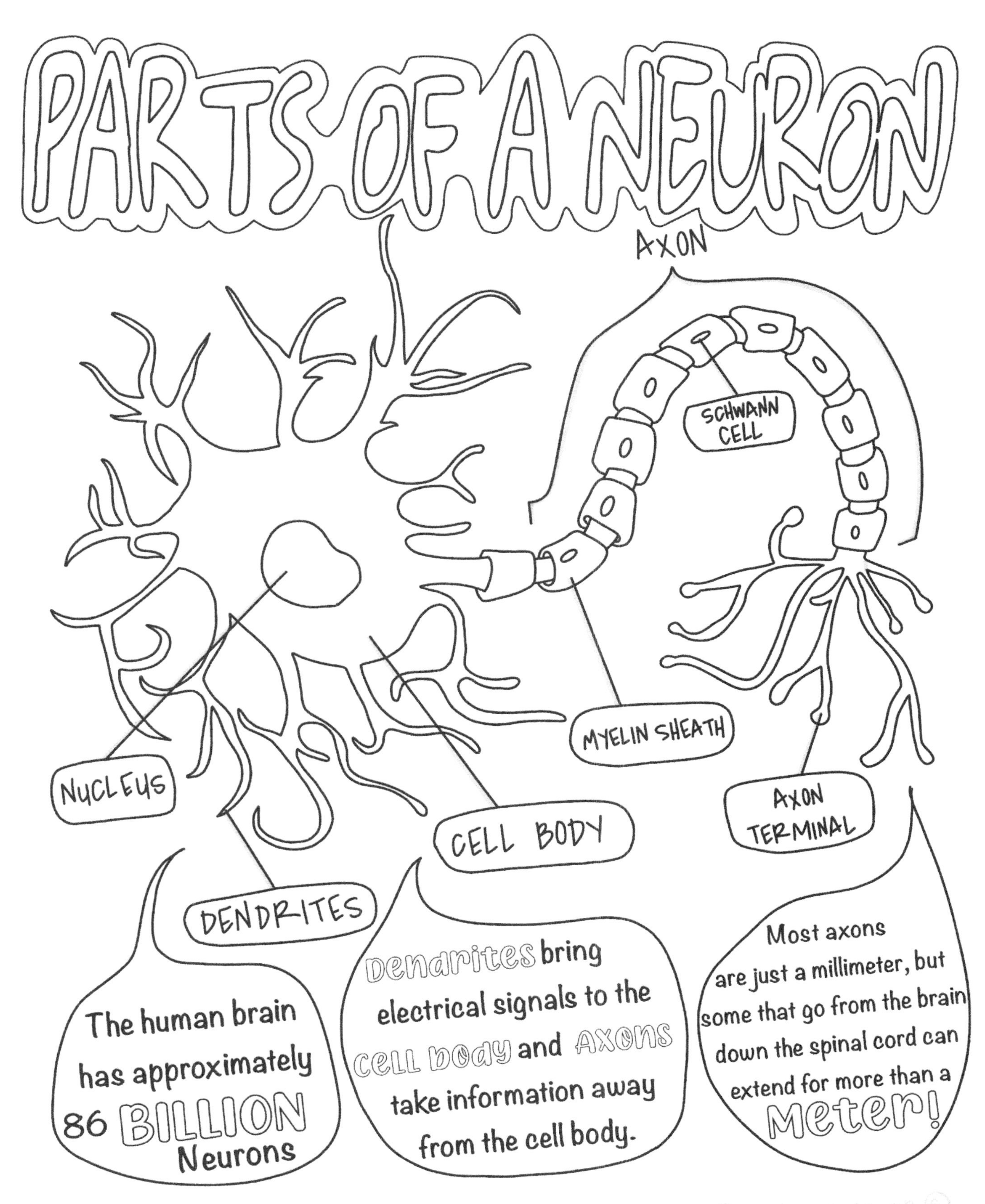

PARTS OF A NEURON
AXON
SCHWANN CELL
MYELIN SHEATH
NUCLEUS
CELL BODY
AXON TERMINAL
DENDRITES
The human brain has approximately 86 BILLION Neurons
Dendrites bring electrical signals to the cell body and Axons take information away from the cell body.
Most axons are just a millimeter, but some that go from the brain down the spinal cord can extend for more than a Meter!

WHAT'S IN YOUR BLOOD?
WHITE BLOOD CELLS
Eosinophil
Neutrophil
Basophil
Monocyte
Lymphocytes
RED BLOOD CELLS
Erythrocytes
PLATELETS
Thrombocytes
↳small, colorless cell fragments that form clots.
Red blood Cells carry oxygen from your lungs to your Tissues, brings carbon dioxide back to your lungs, and makes up almost half of your blood
White blood cells make up less than 1% of your blood, but play the important role in Fighting Infections in your immune system.

WHAT'S A VIRUS?

A VIRUS IS AN INFECTIVE AGENT THAT TYPICALLY CONSISTS OF A NUCLEIC ACID MOLECULE IN A PROTEIN COAT, IS TOO SMALLL TO BE SEEN BY LIGHT MICROSCOPY, AND IS ABLE TO MULTIPLY ONLY WITHIN THE LIVING CELLS OF A HOST.

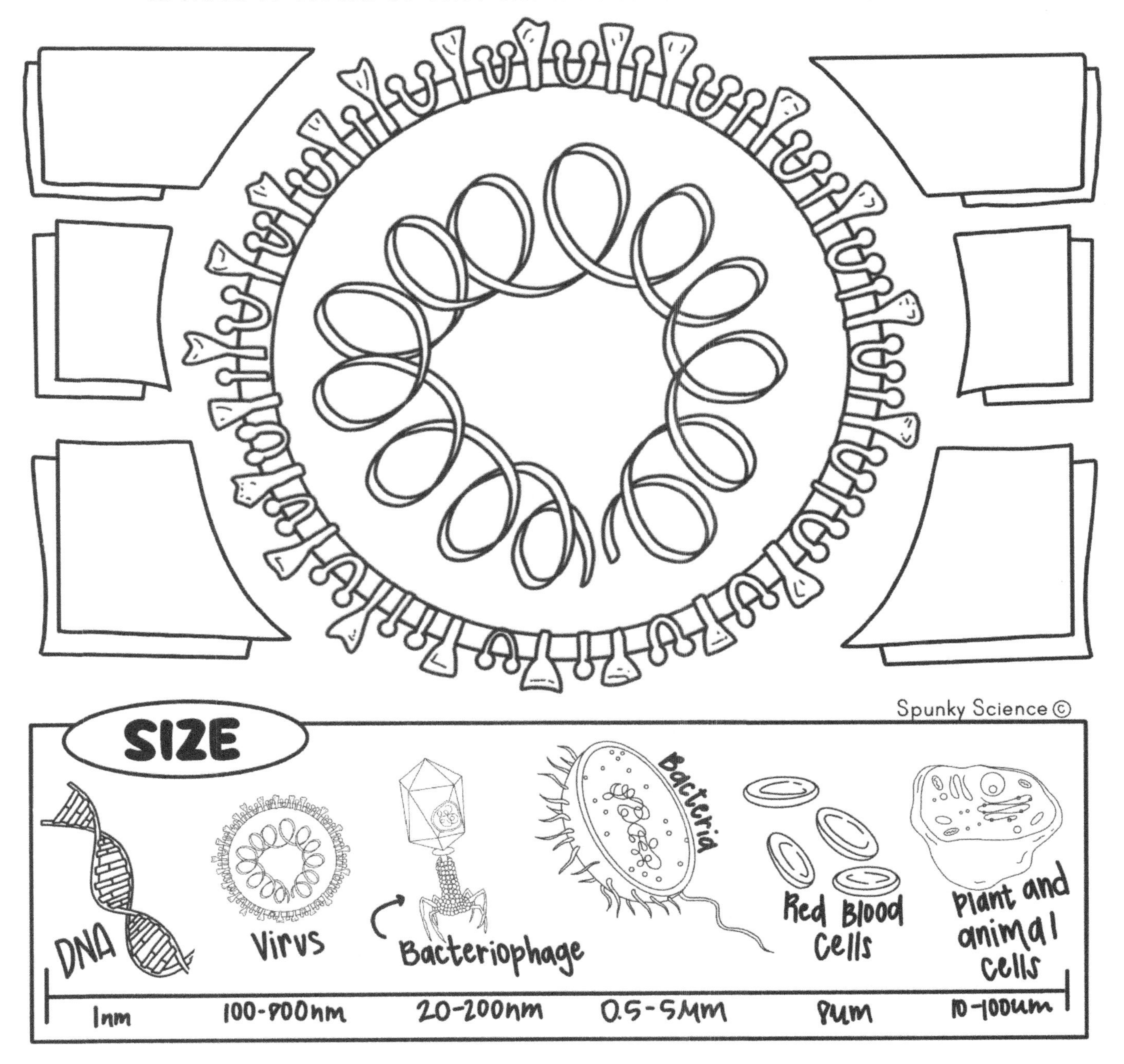

Bacteriophage

A bacteriophage is a virus that infects and replicates within bacteria and archaea. The term was derived from "bacteria" and the Greek word meaning "to devour".

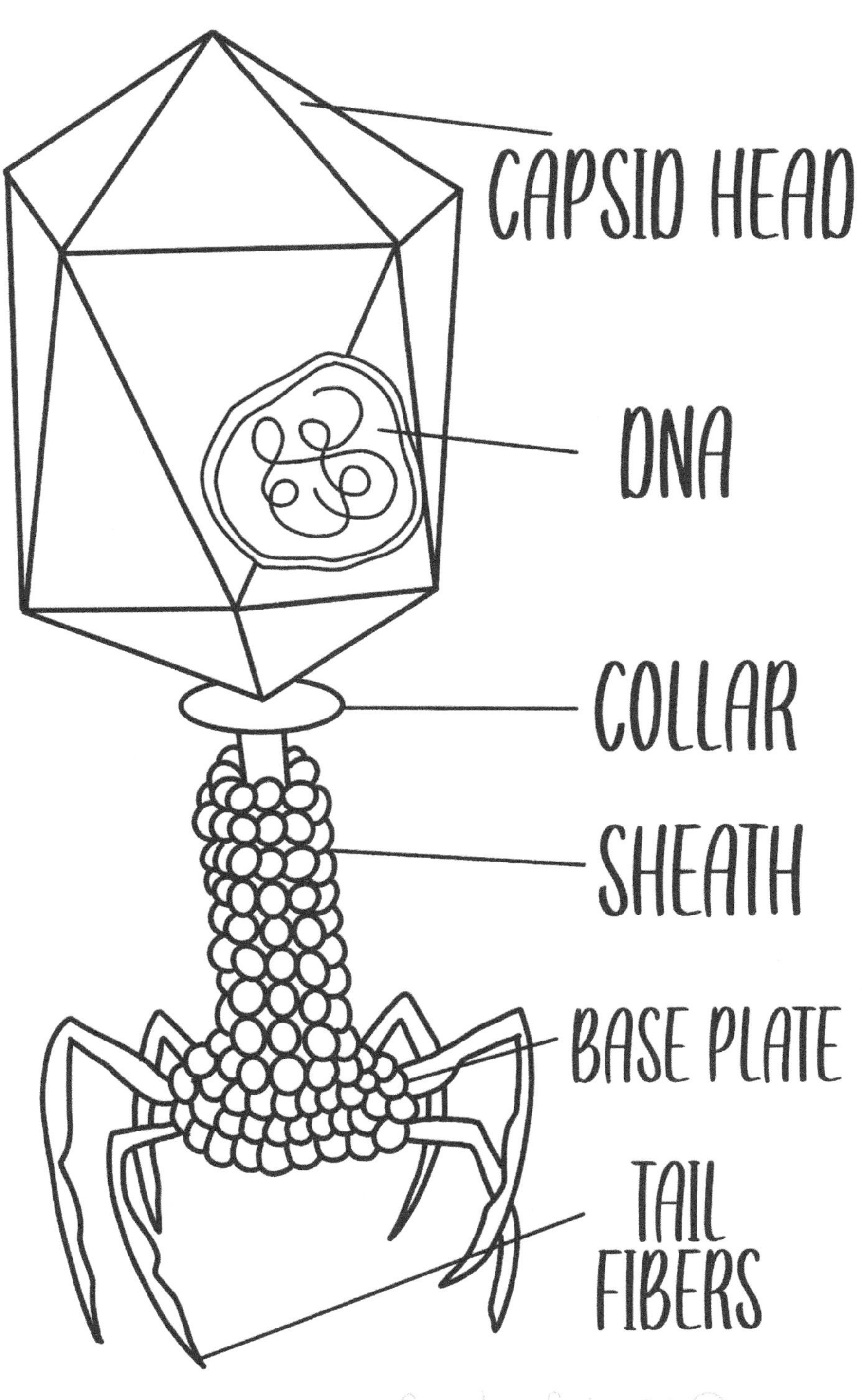

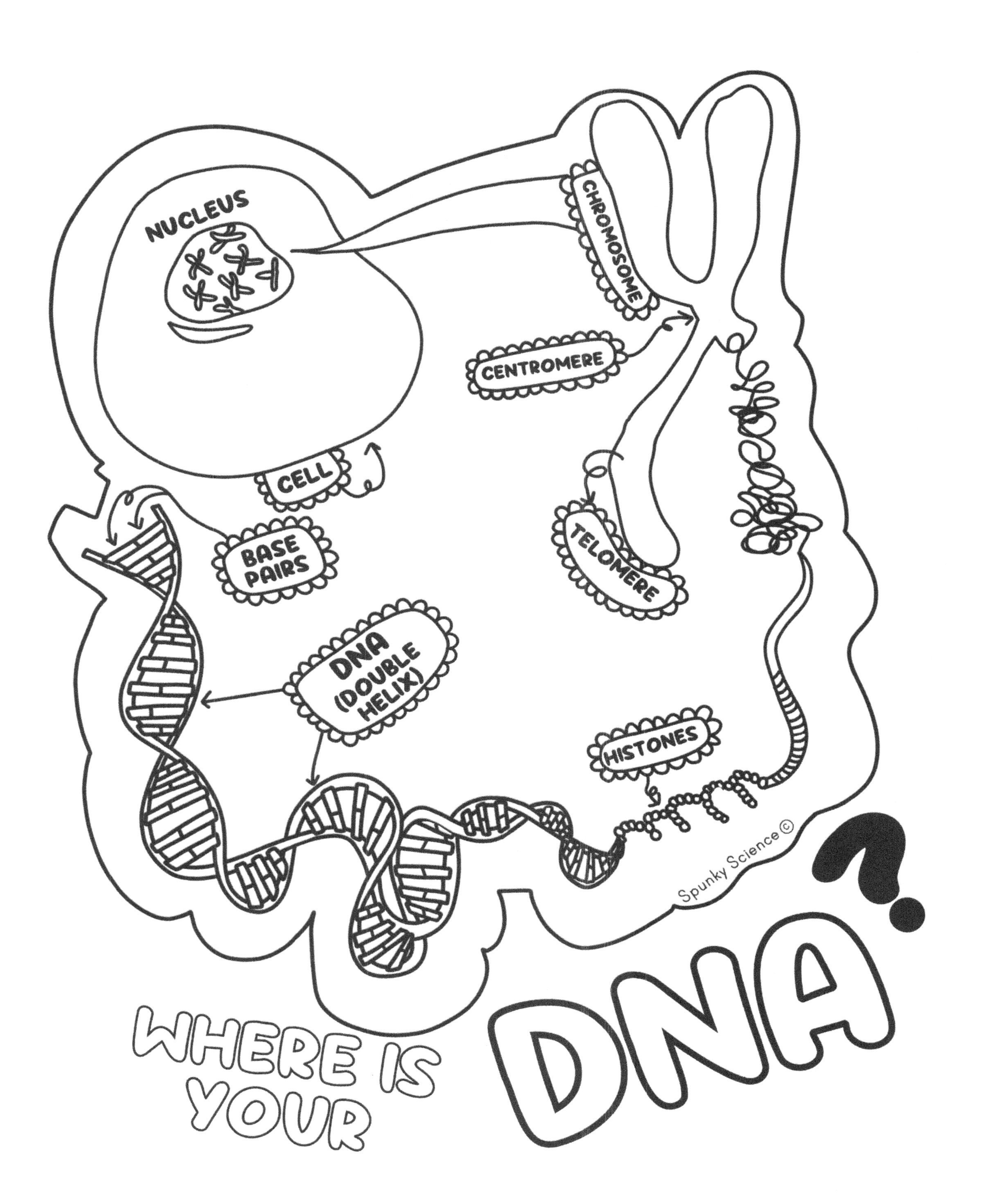
NUCLEUS
CHROMOSOME
CENTROMERE
CELL
BASE PAIRS
TELOMERE
DNA (DOUBLE HELIX)
HISTONES
Spunky Science ©
WHERE IS YOUR DNA?

Stages of

MITOSIS

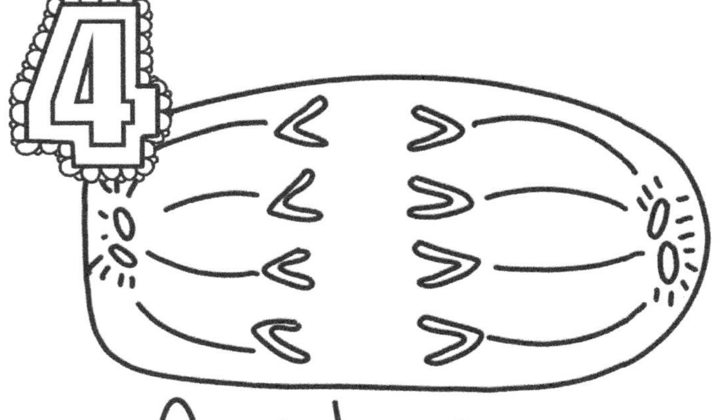

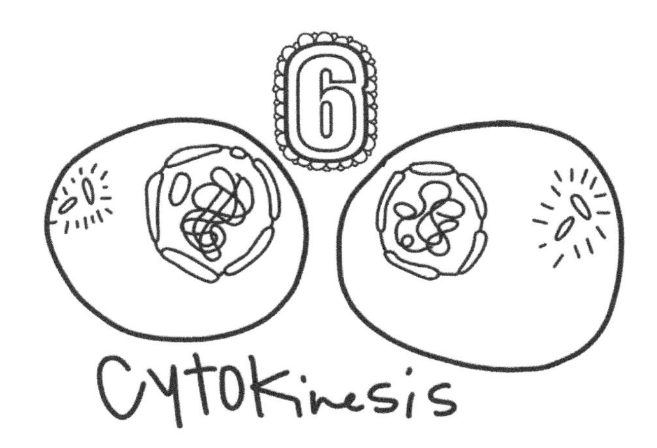

INTERPHASE

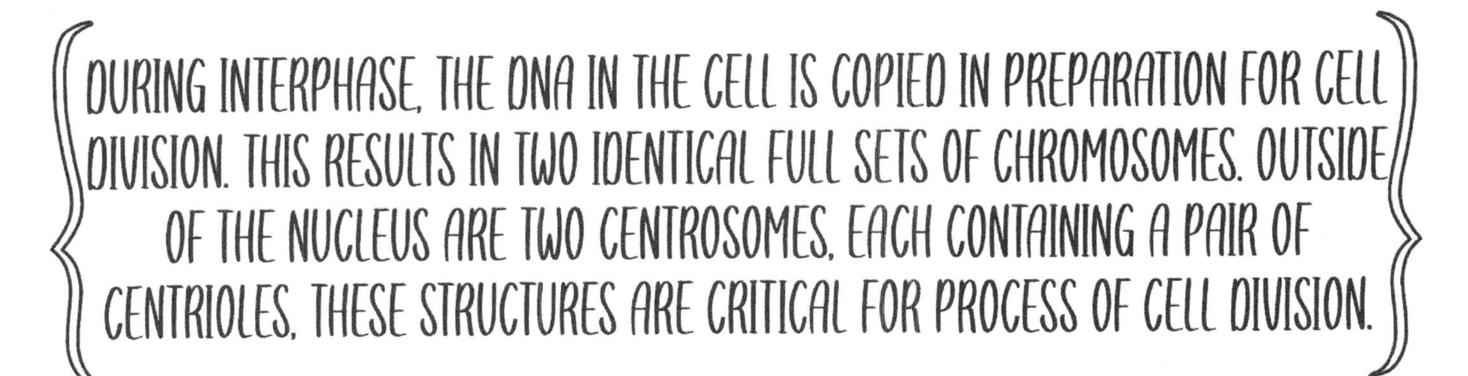

DURING INTERPHASE, THE DNA IN THE CELL IS COPIED IN PREPARATION FOR CELL DIVISION. THIS RESULTS IN TWO IDENTICAL FULL SETS OF CHROMOSOMES. OUTSIDE OF THE NUCLEUS ARE TWO CENTROSOMES, EACH CONTAINING A PAIR OF CENTRIOLES. THESE STRUCTURES ARE CRITICAL FOR PROCESS OF CELL DIVISION.

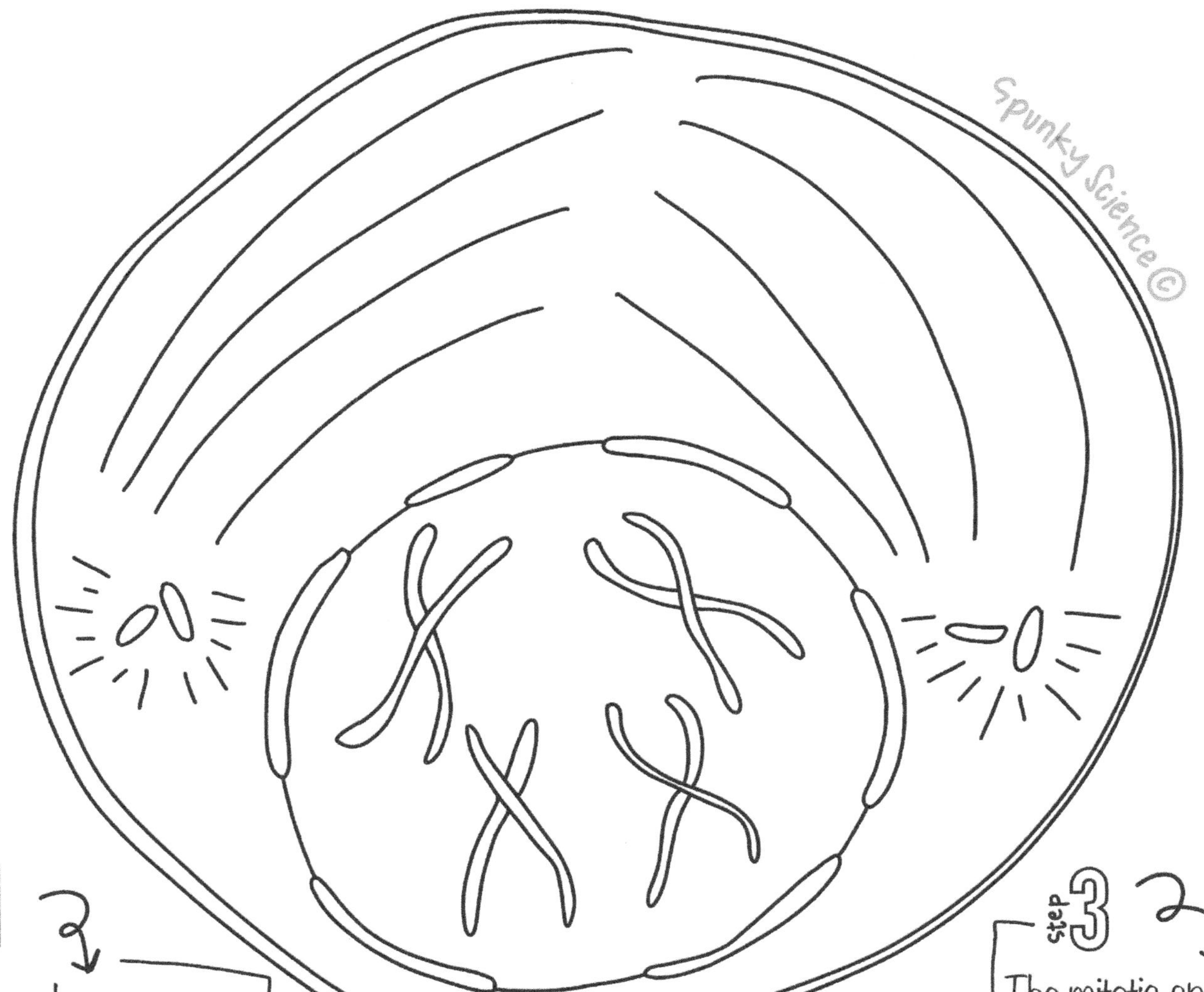

step 1

The chromosomes condense into x-shaped structures that can be easily seen under a microscope. Each chromosome is composed of two sister chromatids containing identical genetic information.

step 2

The chromosomes pair up so that both copies of chromosome 1 are together, both copies of chromosome 2 are together, and so on. At the end of prophase the membrane around the nucleus dissolves away releasing the chromosomes.

step 3

The mitotic spindle consisting of the microtubles and other proteins, extends across the cell between the centrioles as they move to opposite poles of the cell.

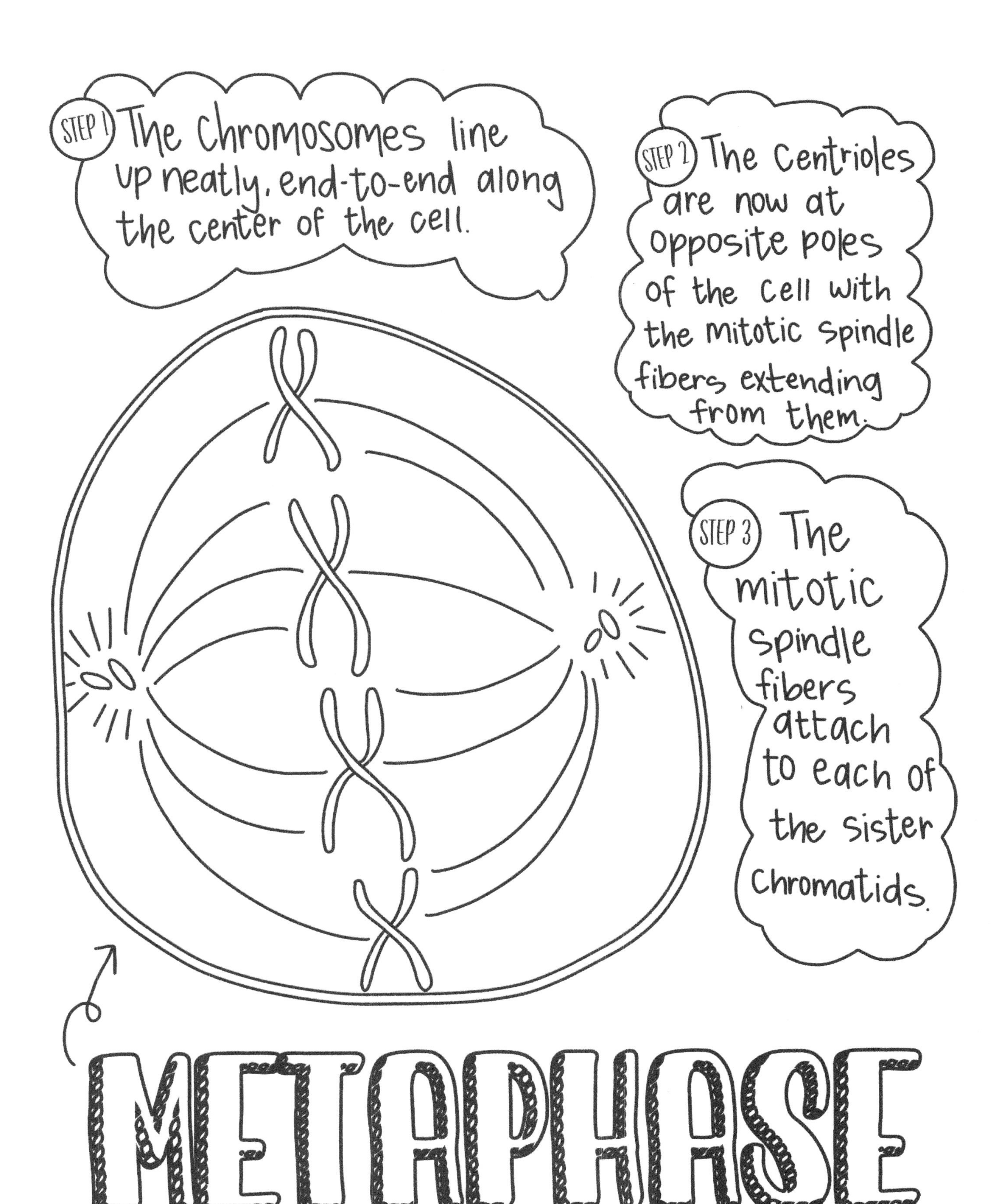
STEP 1 The Chromosomes line up neatly, end-to-end along the center of the cell.
STEP 2 The Centrioles are now at opposite poles of the cell with the mitotic spindle fibers extending from them.
STEP 3 The mitotic spindle fibers attach to each of the sister Chromatids.

METAPHASE

THE SISTER CHROMATIDS ARE PULLED APART BY THE MITOTIC SPINDLE WHICH PULLS ONE CHROMATID TO ONE POLE AND THE OTHER TO THE OPPOSITE POLE.

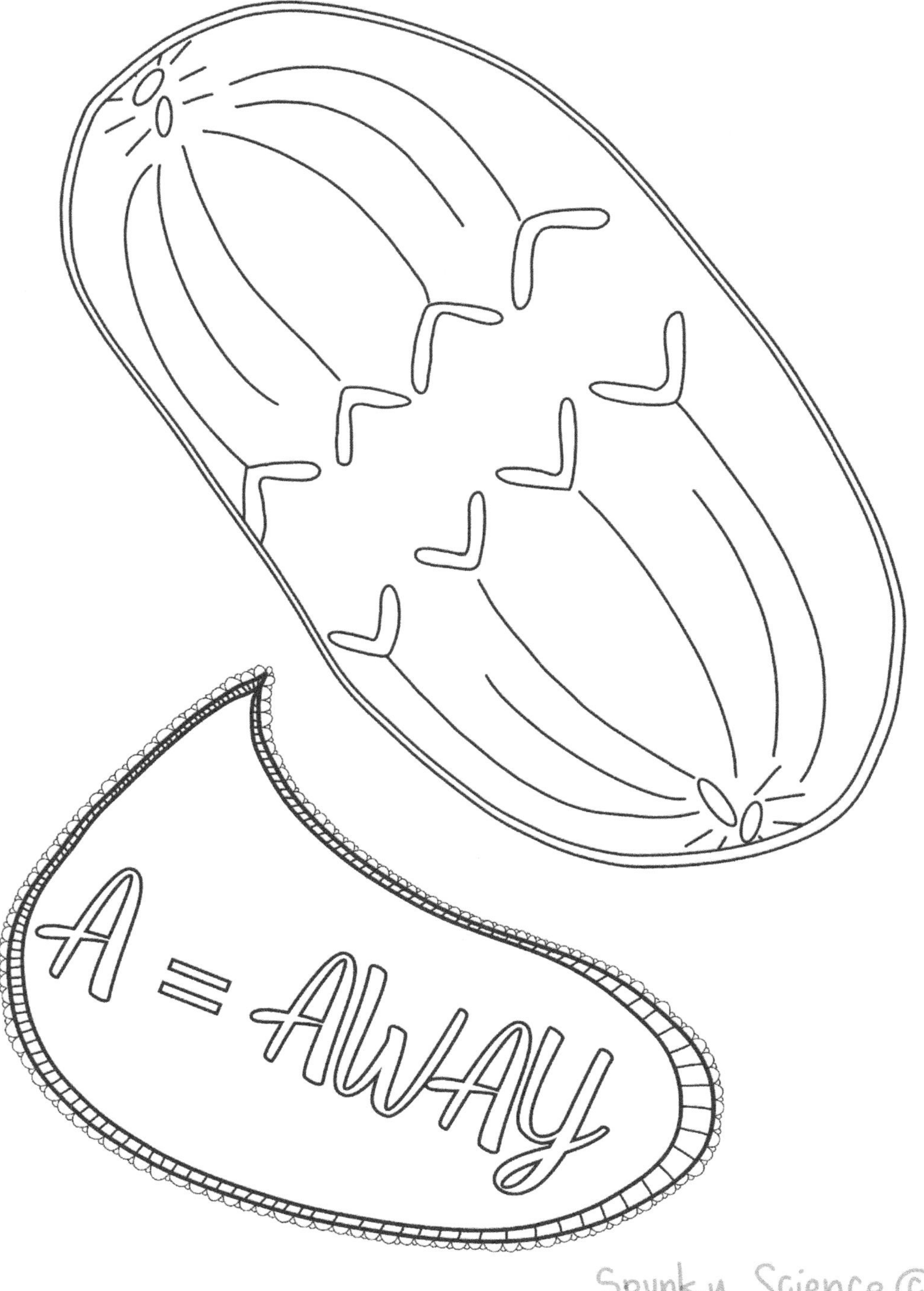

TELOPHASE

First

AT EACH POLE OF THE CELL A FULL SET OF CHROMOSOMES GATHER TOGETHER.

Second

A MEMBRANE FORMS AROUND EACH SET OF CHROMOSOMES TO CREATE TWO NEW NUCLEI.

CYTOKINESIS

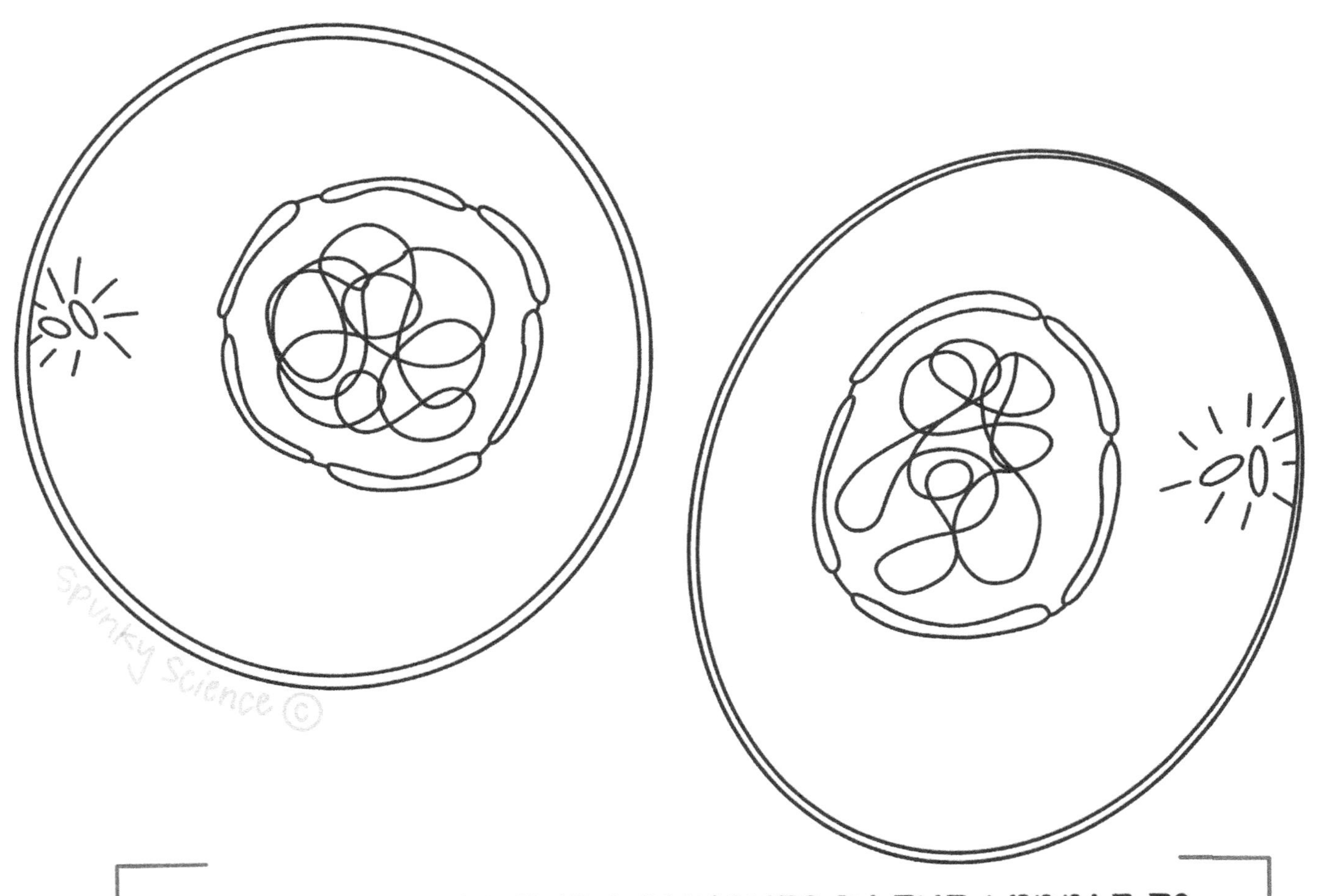

THE SINGLE CELL THEN PUNCHES IN THE MIDDLE TO FORM TWO SEPARATE DAUGHTER CELLS EACH CONTAINING A FULL SET OF CHROMOSOMES WITHIN A NUCLEUS.

GOLGI APPARATUS
Packages proteins into membrane bound vesicles inside the cell before the vesicles are sent to their destination.
it is located in the cytoplasm next to the endoplasmic reticulum and near the cell nucleus.
Cell's " Post Office"
©Spunky Science

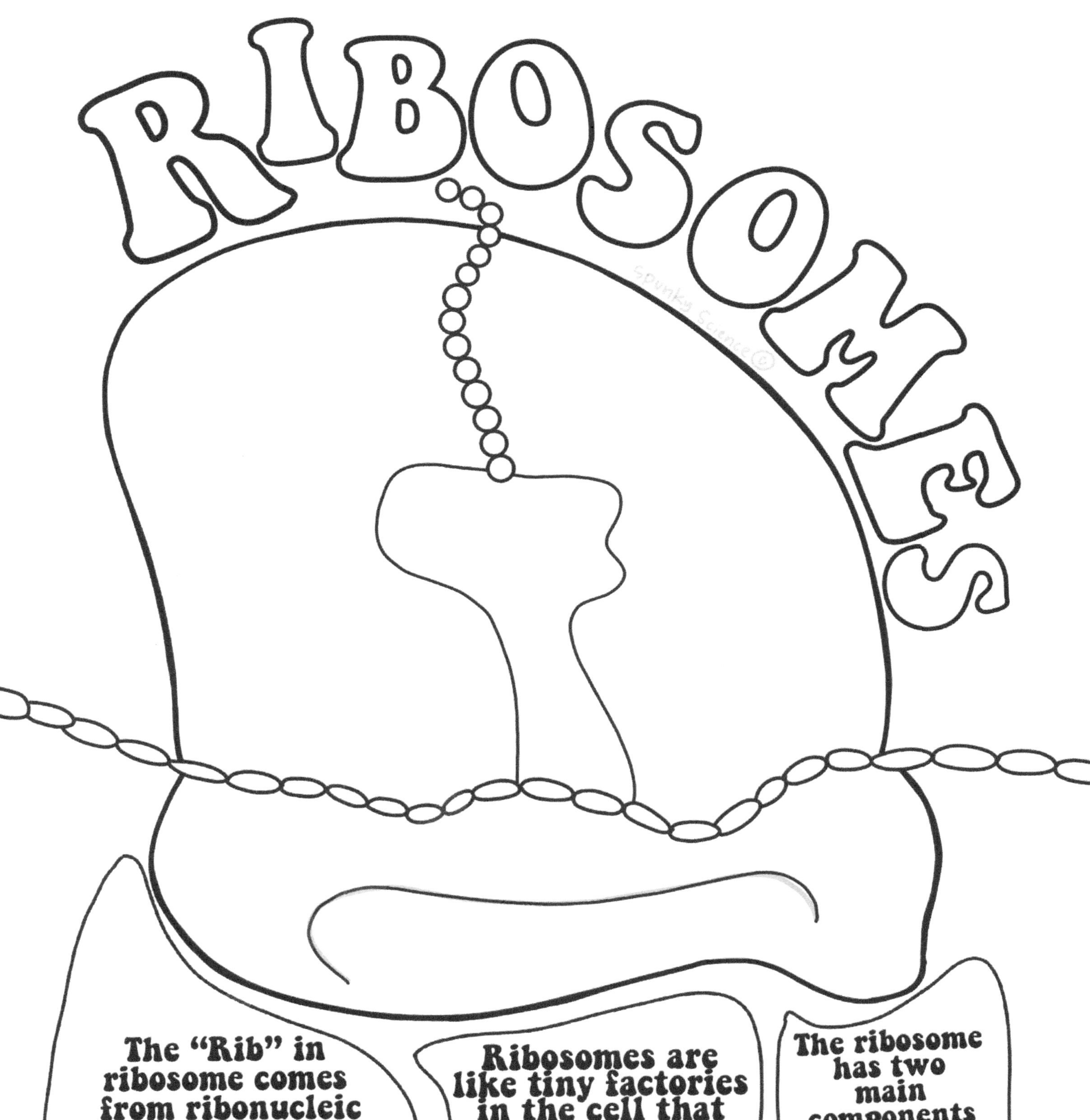

The "Rib" in ribosome comes from ribonucleic acid (RNA) which provides instructions on making proteins.

Ribosomes are like tiny factories in the cell that make proteins and perform functions for the cell's operations.

The ribosome has two main components called the large subunit and the small subunit.

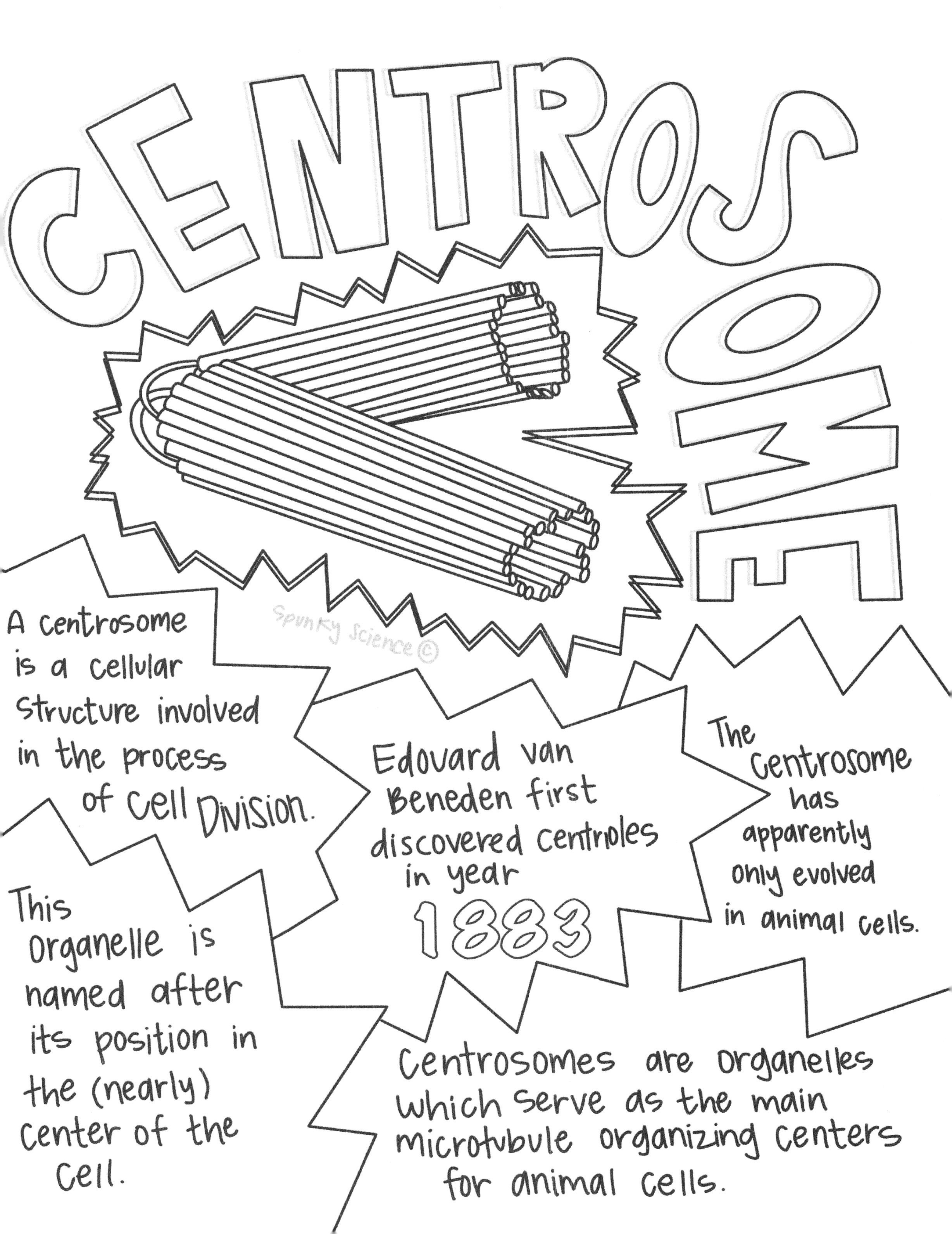
CENTROSOME
Spunky Science ©
A centrosome is a cellular structure involved in the process of cell Division.
Edovard van Beneden first discovered centrioles in year 1883
The Centrosome has apparently only evolved in animal cells.
This organelle is named after its position in the (nearly) center of the cell.
Centrosomes are organelles which serve as the main microtubule organizing centers for animal cells.

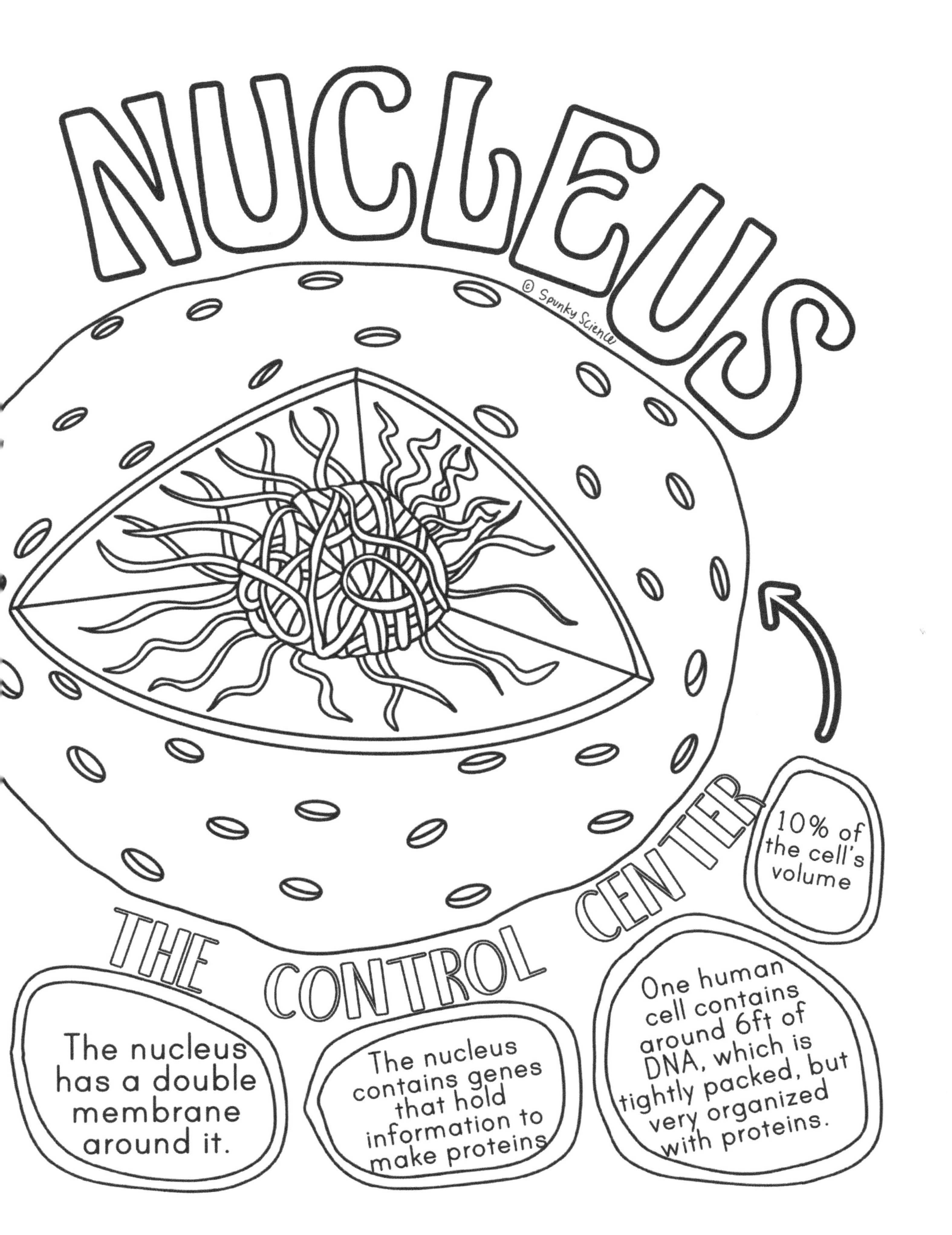
NUCLEUS
© Spunky Science
THE CONTROL CENTER
10% of the cell's volume
The nucleus has a double membrane around it.
The nucleus contains genes that hold information to make proteins
One human cell contains around 6ft of DNA, which is tightly packed, but very organized with proteins.

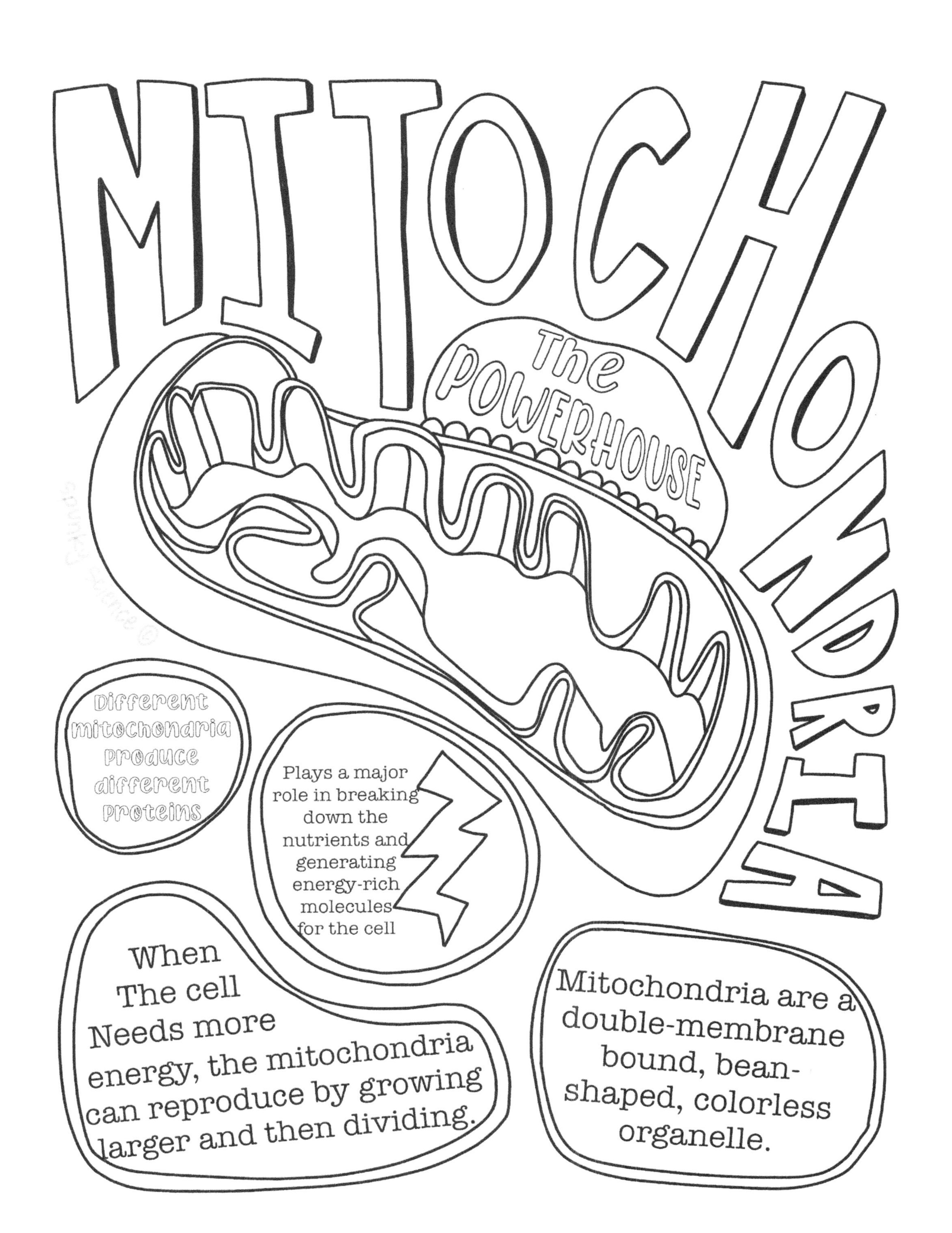
MITOCHONDRIA
The POWERHOUSE
Different mitochondria produce different proteins
Plays a major role in breaking down the nutrients and generating energy-rich molecules for the cell
When The cell Needs more energy, the mitochondria can reproduce by growing larger and then dividing.
Mitochondria are a double-membrane bound, bean-shaped, colorless organelle.

ENDOPLASMIC
RETICULUM
Rough ER has RIBOSOMES attached to it.
In Eukaryotic cells, almost 50% of the surface consists of ER.
ER is short for Endoplasmic Reticulum
Transportation system and protein folding
Spunky Science©

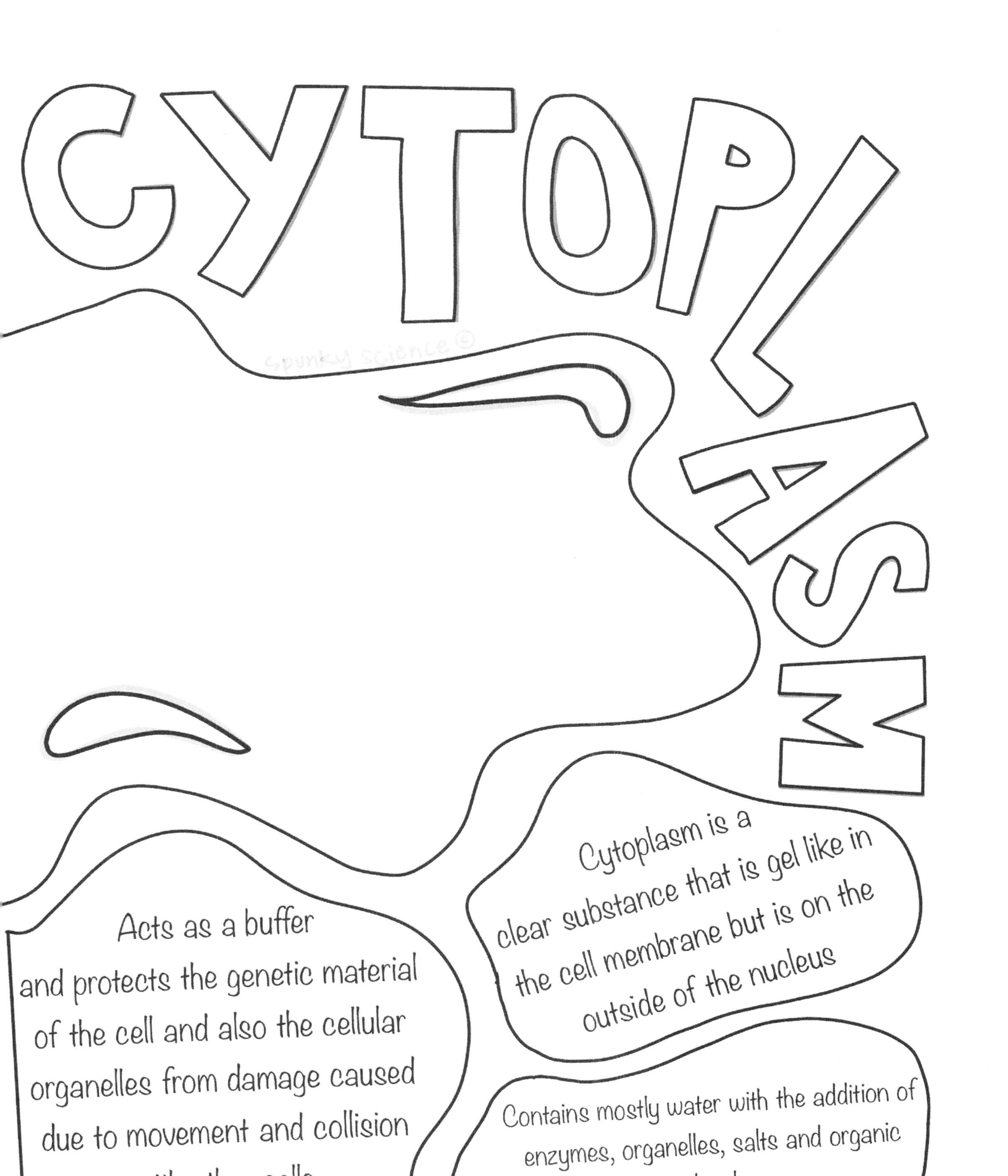

Acts as a buffer and protects the genetic material of the cell and also the cellular organelles from damage caused due to movement and collision with other cells.

Cytoplasm is a clear substance that is gel like in the cell membrane but is on the outside of the nucleus

Contains mostly water with the addition of enzymes, organelles, salts and organic molecules.

Central VACUOLE

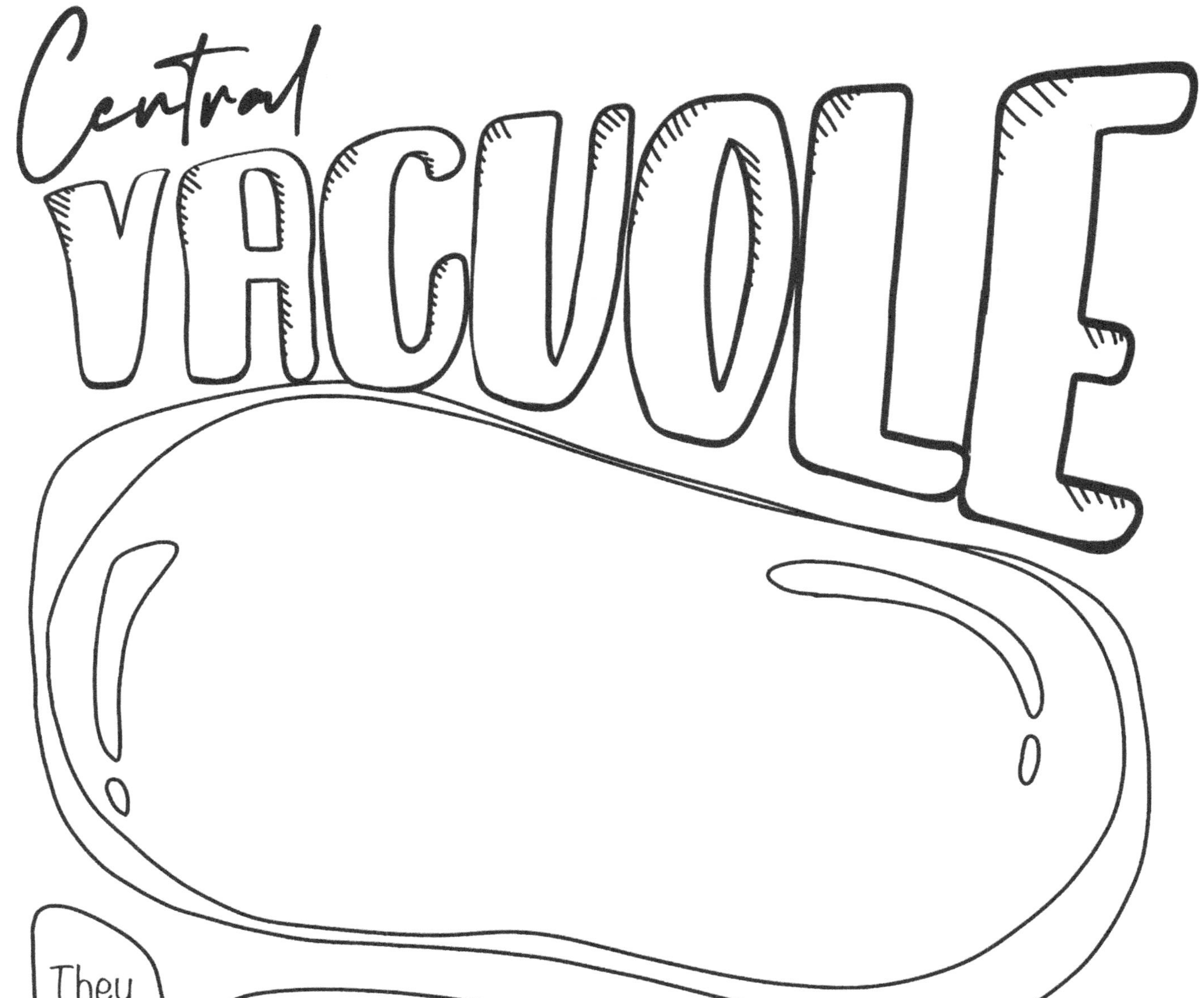

They are found I n both animal and plant cells, but are much larger in plant cells.

Holds WATER, FOOD, and waste for the cell.

As TURGOR pressure builds, it pushes out against the sides of the cell and gives it structure.

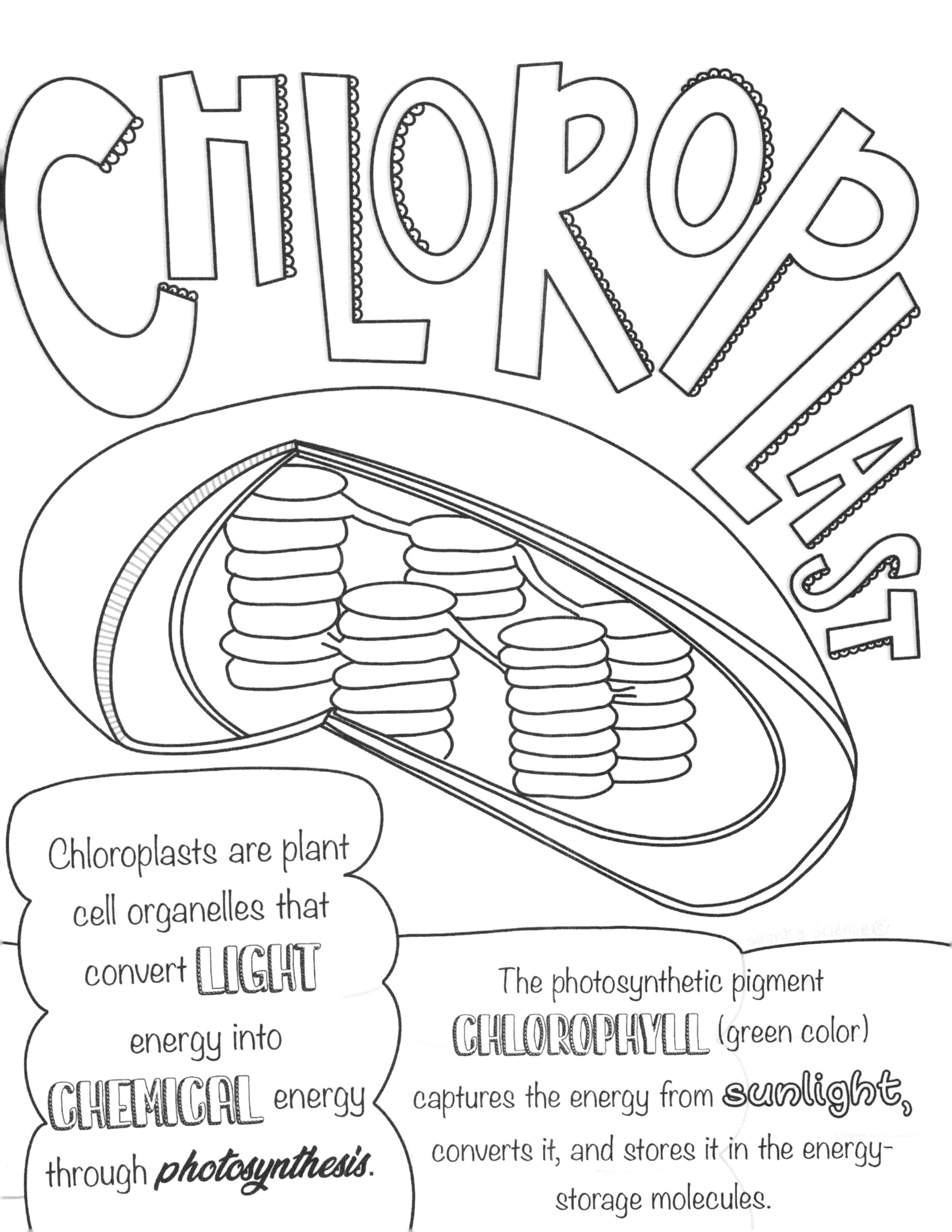
CHLOROPLAST
Chloroplasts are plant cell organelles that convert LIGHT energy into CHEMICAL energy through photosynthesis.
The photosynthetic pigment CHLOROPHYLL (green color) captures the energy from sunlight, converts it, and stores it in the energy-storage molecules.

3 TYPES OF MUSCLE TISSUES

- Voluntary control of the nervous system
- Attached to bones by tendons

- Located in the walls of hollow visceral organs except the heart
- Involuntary control of the nervous system

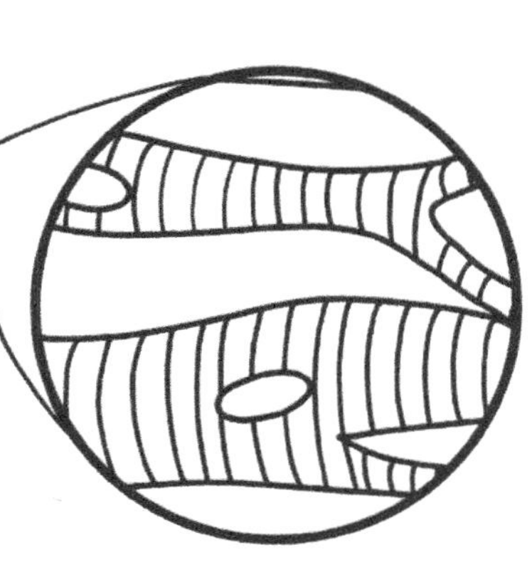

- In the walls of the heart
- Striated
- Under involuntary control

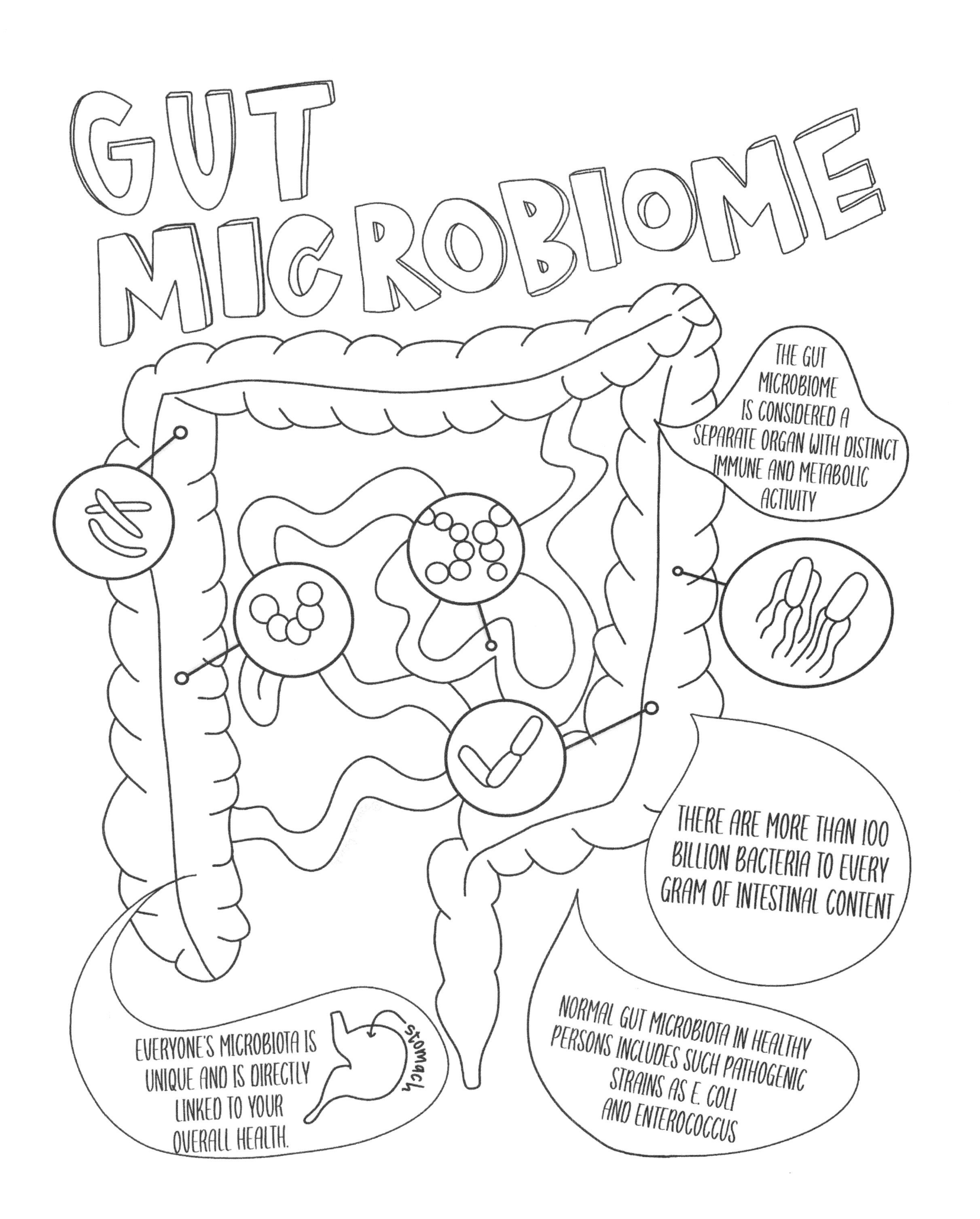
GUT MICROBIOME
THE GUT MICROBIOME IS CONSIDERED A SEPARATE ORGAN WITH DISTINCT IMMUNE AND METABOLIC ACTIVITY
THERE ARE MORE THAN 100 BILLION BACTERIA TO EVERY GRAM OF INTESTINAL CONTENT
EVERYONE'S MICROBIOTA IS UNIQUE AND IS DIRECTLY LINKED TO YOUR OVERALL HEALTH.
stomach
NORMAL GUT MICROBIOTA IN HEALTHY PERSONS INCLUDES SUCH PATHOGENIC STRAINS AS E. COLI AND ENTEROCOCCUS

YOUR STOMACH
spunky science ©
It contains hydrochloride acid, which helps to kill off bacteria and viruses that may enter the foods you eat.
J shaped organ
When you blush, the lining of your stomach also turns red!
The stomach is the widest part of the digestive system.
Enzymes in your digestive system are what separate food into different nutrients that your body needs.
The stomach can hold a bit more than a liter of food at once.

Thyroid

THE THYROID GLAND LOOKS LIKE A *butterfly* AND IS LOCATED AT THE FRONT OF THE NECK BELOW THE ADAM'S APPLE.

ALMOST EVERY CELL IN YOUR BODY NEEDS THYROID HORMONES

SOMETIMES THE THYROID IS *hyperactive* AND PRODUCES TOO MUCH THYROID HORMONE.

TO WORK PROPERLY, THE THYROID NEEDS IODINE.

AMONG MANY OTHER THINGS, THYROID HORMONES HELP **control** HOW FAST YOU BURN CALORIES, HOW FAST YOUR HEART BEATS, AND YOUR BODY TEMPERATURE.

Anxiety AND insomnia CAN BE SIGNS OF AN OVERACTIVE THYROID

THE THYROID IS UNDER THE *control* OF A PEANUT SHAPED GLAD IN THE BRAIN CALLED THE **PITUITARY** GLAND.

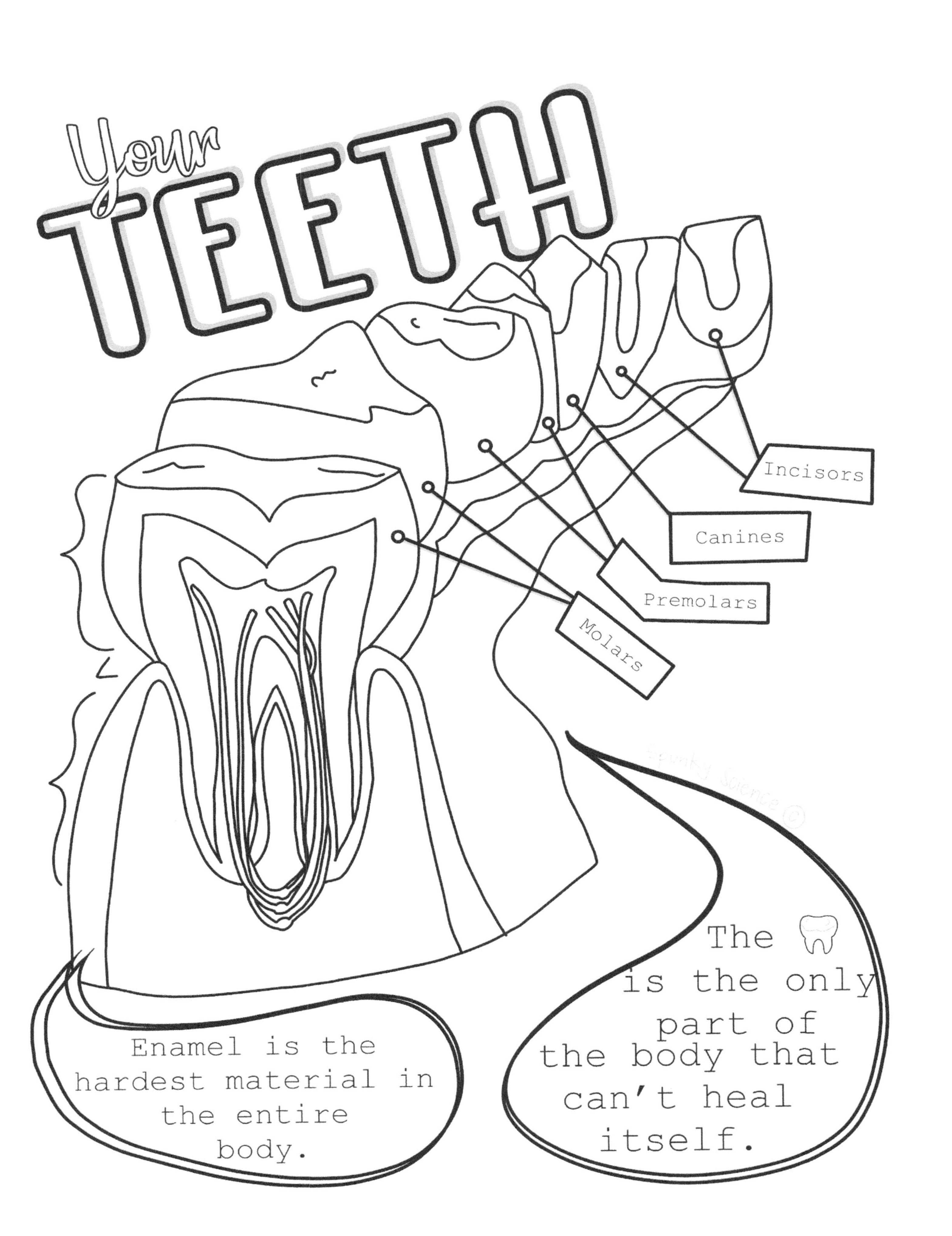
Your
TEETH
Incisors
Canines
Premolars
Molars
Enamel is the hardest material in the entire body.
The is the only part of the body that can’t heal itself.

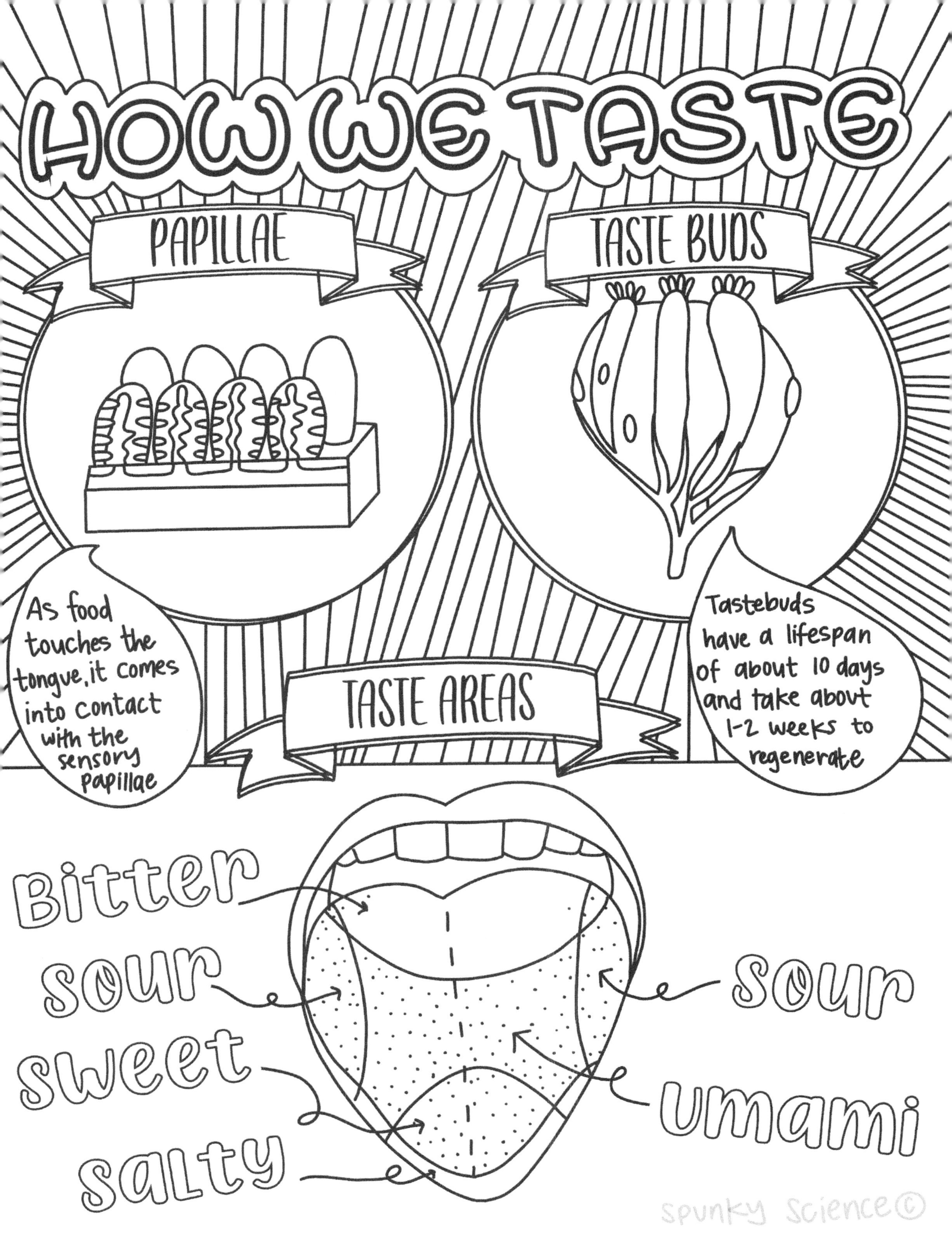
HOW WE TASTE
PAPILLAE
TASTE BUDS
As food touches the tongue, it comes into contact with the sensory papillae
Tastebuds have a lifespan of about 10 days and take about 1-2 weeks to regenerate
TASTE AREAS
Bitter
sour
sweet
salty
sour
umami
spunky science©

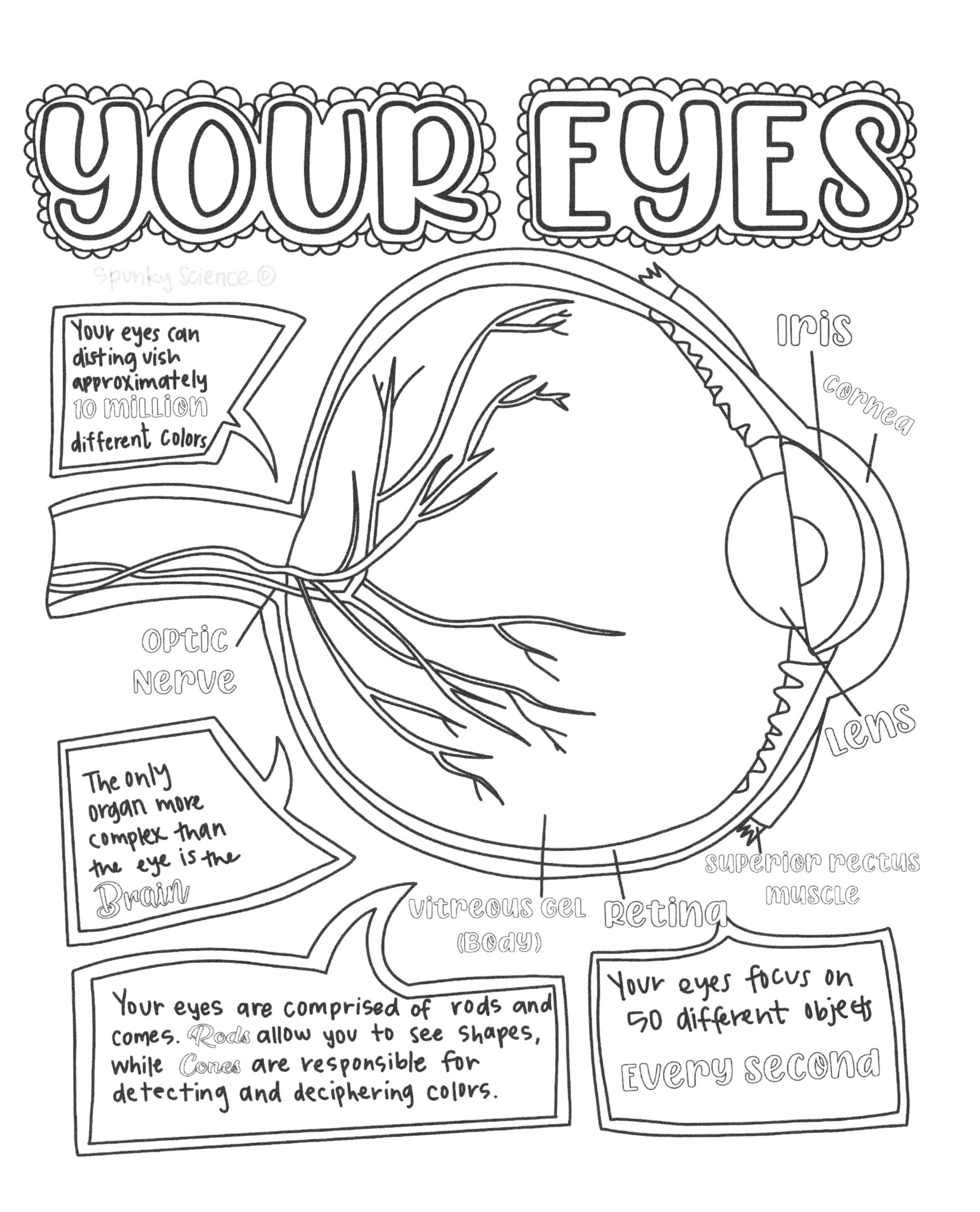

YOUR EYES
Spunky Science ©
Your eyes can disting vish approximately 10 MILLION different colors
Iris
Cornea
Optic Nerve
Lens
The only organ more complex than the eye is the Brain
Superior rectus muscle
Vitreous Gel (Body)
Retina
Your eyes are comprised of rods and comes. Rods allow you to see shapes, while Cones are responsible for detecting and deciphering colors.
Your eyes focus on 50 different objects Every second

BODY SYSTEMS

WORD FIND

s t a l c n w i n r c r t o p s c a y a a

b u l t a r e o r e o e n u m m i b r f m

m s c i d c a b d p j s d p t n r g a n i

e a o s f w t f d r d p k l a z c d t r b

u r p s w e n t l o z i e s t n u e n k m

j h e u i x p e o d c r n r e l l m e c e

k g r e n d t f p u p a r q d m a s m i n

n d t s u c y l e c b t f w t u t a u e a

e e o p v s z a b t y o g x y s o a g d m

r g a w n o k x m i a r e q w c r o e n c

v a f g u n q e r v s y n r e u y r t a b

o m v r l o p s l e s z l u c l g e n s t

u r i n a r y b u e y c e s l a i d i v i

s k l e t l e c r i t r i e n r o m m u d

a d i g e s t i v e n a e n d o c r i n e

c r s d l m a d u i n n l r e w o s a i e

WORDBANK

- immune
- skeletal
- muscular
- digestive
- circulatory
- integumentary
- nervous
- respiratory
- urinary
- reproductive
- endocrine
- tissues

The Digestive System
spunky science ©
It takes 24 TO 48 HOURS to fully digest food.
MOUTH
SALIVARY GLANDS
PHARYNX
ESOPHAGUS
The average person produces 2 pints of saliva every day
Can be up to 30 feet in length in adults
LIVER
STOMACH
GALL BLADDER
PANCREAS
The gut-brain axis is the close bond that exists between the digestive system and your brain.
The second part of your small intestine is called the jejunum.
LARGE INTESTINE
APPENDIX
SMALL INTESTINE
RECTUM
Digestion is the breakdown of food into small molecules, which are then absorbed into the body.
ANUS
Stomach growling is called borborygmic and happens all the time, but it is just louder when your stomach is empty because there isn't food to muffle the sound.

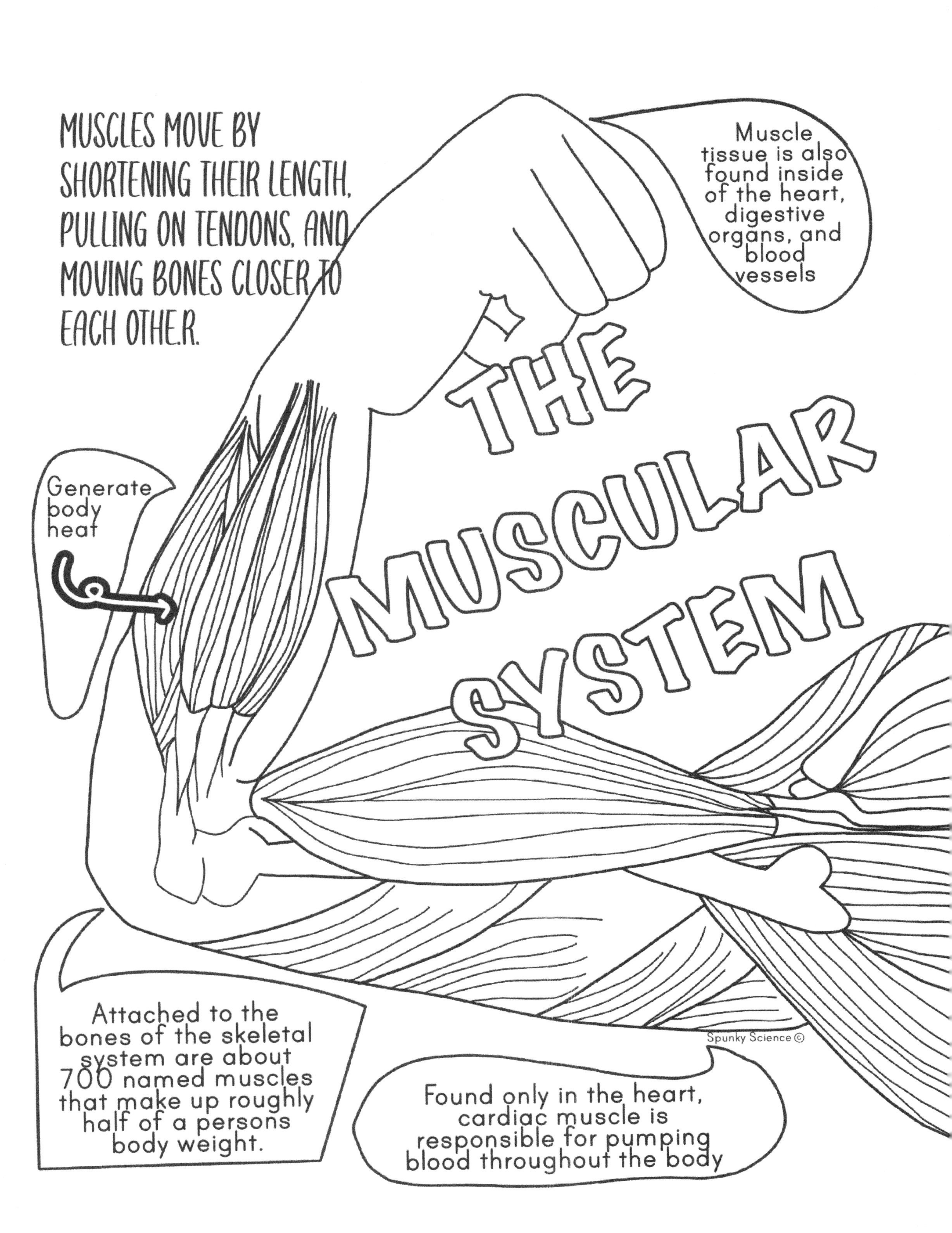
MUSCLES MOVE BY SHORTENING THEIR LENGTH, PULLING ON TENDONS, AND MOVING BONES CLOSER TO EACH OTHER.R.
Muscle tissue is also found inside of the heart, digestive organs, and blood vessels
THE MUSCULAR SYSTEM
Generate body heat
Attached to the bones of the skeletal system are about 700 named muscles that make up roughly half of a persons body weight.
Spunky Science ©
Found only in the heart, cardiac muscle is responsible for pumping blood throughout the body

THE BRAIN AND THE NERVOUS SYSTEM

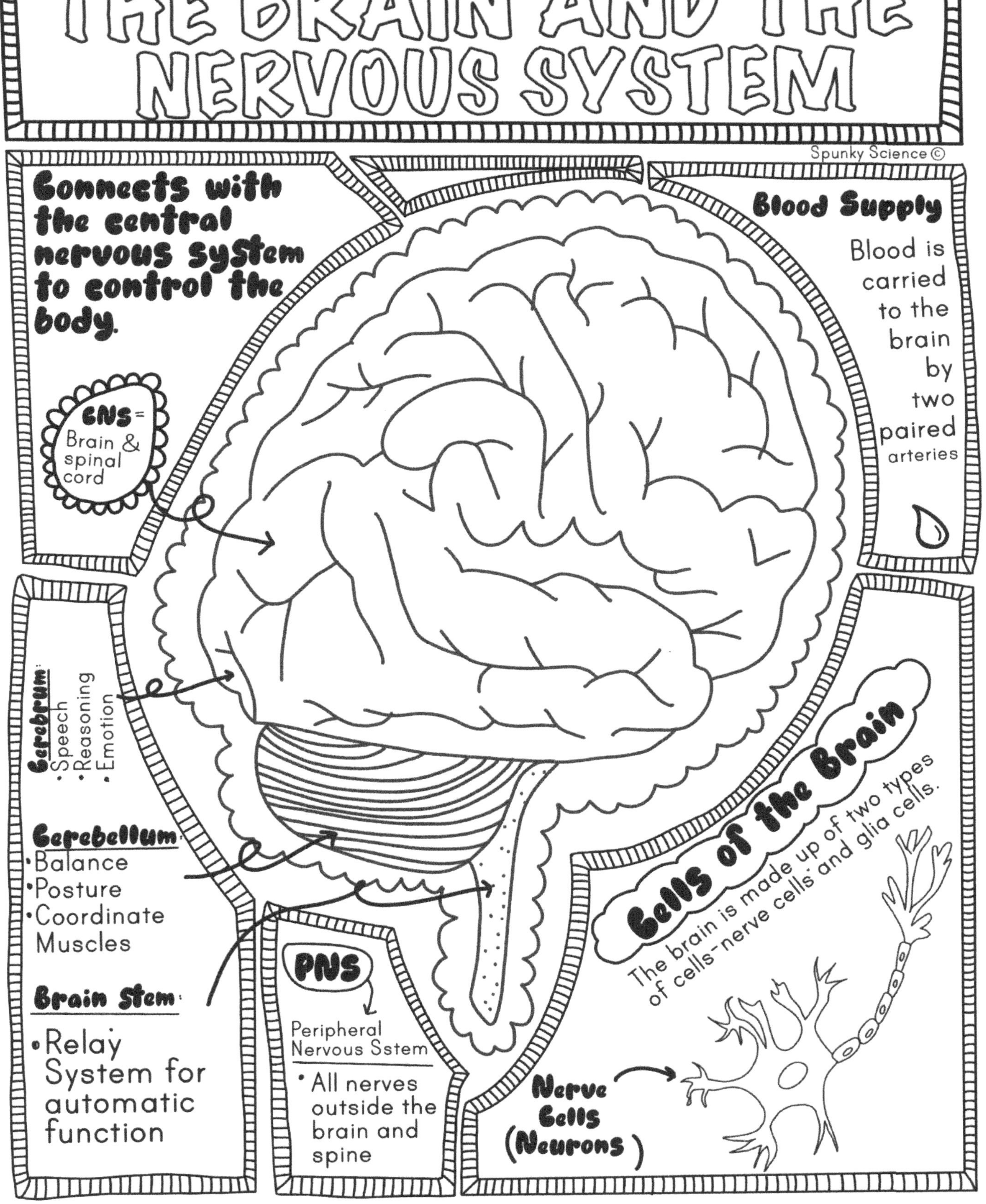

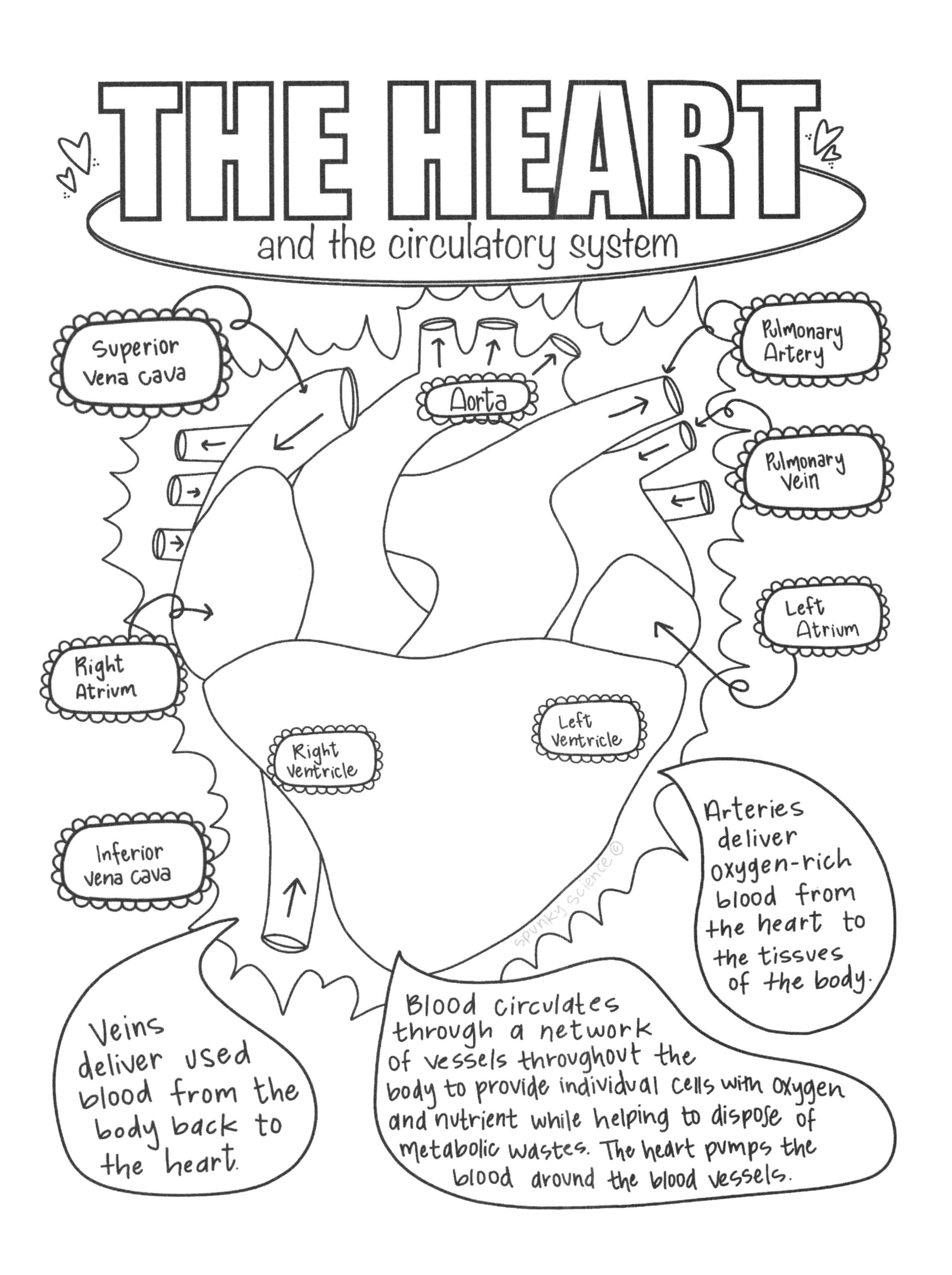
THE HEART
and the circulatory system
Superior Vena cava
Aorta
Pulmonary Artery
Pulmonary Vein
Left Atrium
Right Atrium
Left Ventricle
Right Ventricle
Inferior vena cava
Arteries deliver oxygen-rich blood from the heart to the tissues of the body.
Veins deliver used blood from the body back to the heart.
Blood circulates through a network of vessels throughout the body to provide individual cells with oxygen and nutrient while helping to dispose of metabolic wastes. The heart pumps the blood around the blood vessels.
spunky science ©

EXCRETORY SYSTEM

The systems that excrete wastes from the body. For example, the system of organs that regulates the amount of water in the body and filters and eliminates from the blood the wastes produced by metabolism. The principal organs of the excretory system are the kidneys, ureters, urethra, and urinary bladder.

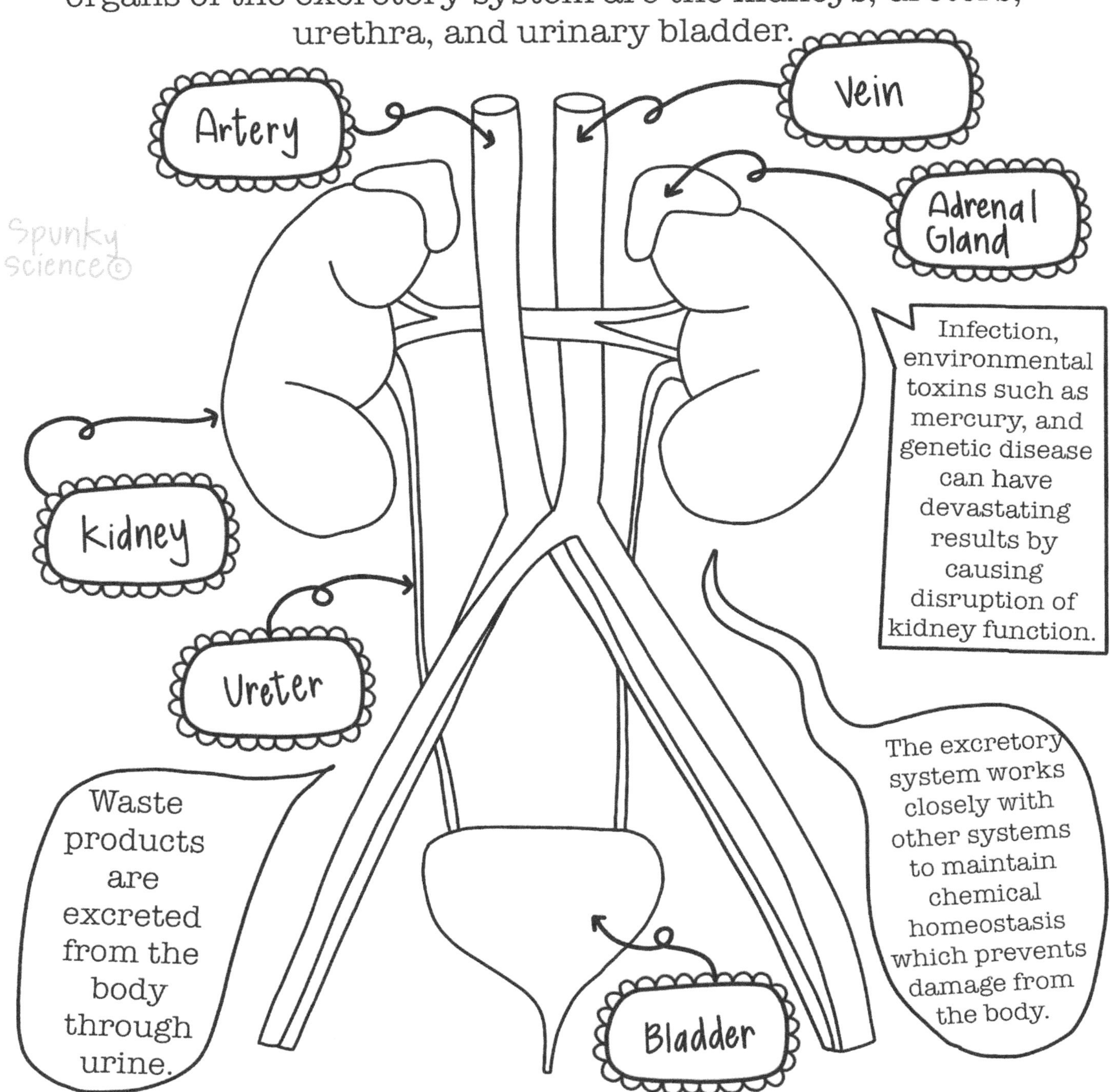

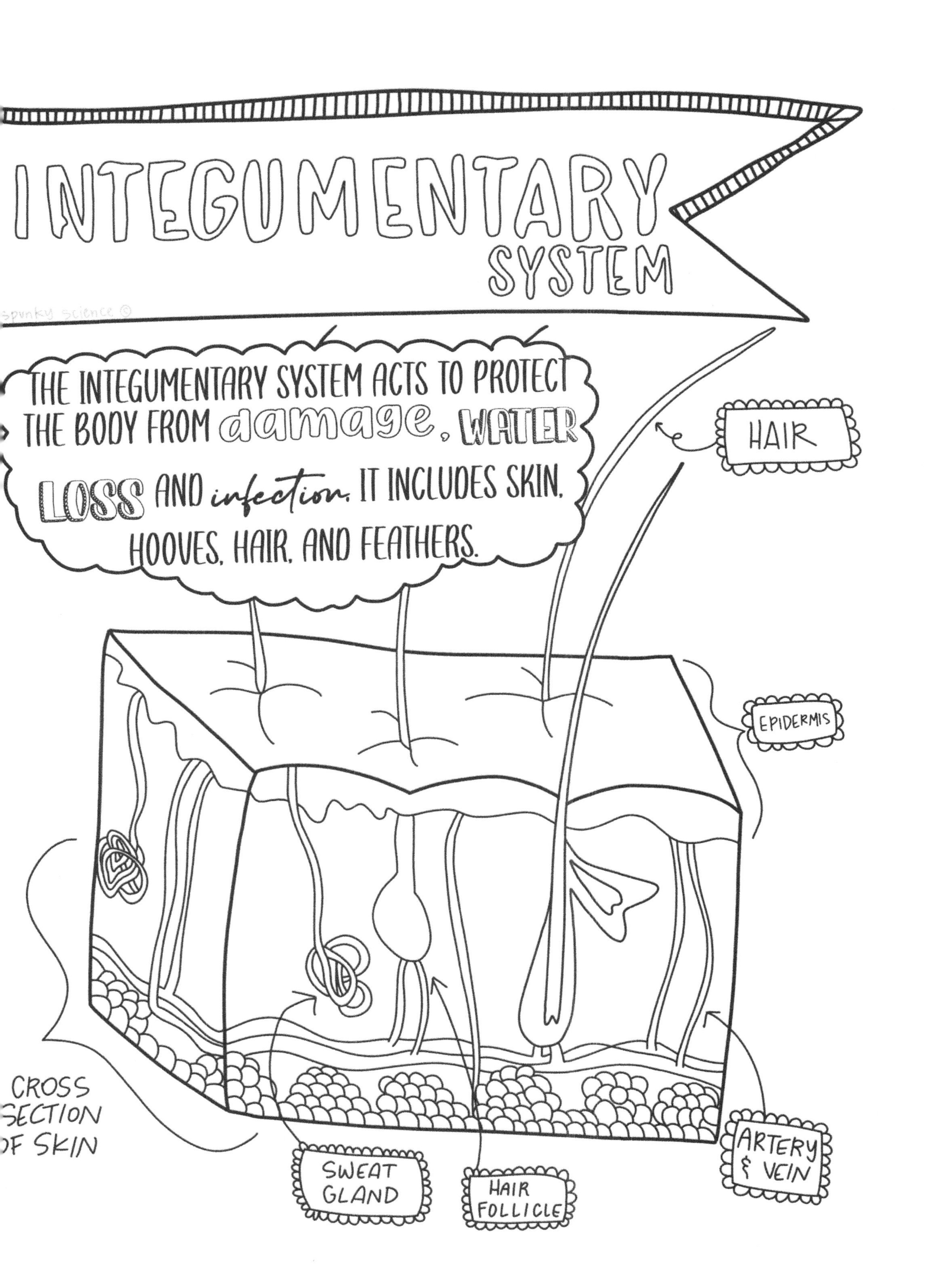
INTEGUMENTARY
SYSTEM
spunky science ©
THE INTEGUMENTARY SYSTEM ACTS TO PROTECT THE BODY FROM damage, WATER LOSS AND infection. IT INCLUDES SKIN, HOOVES, HAIR, AND FEATHERS.
HAIR
EPIDERMIS
CROSS SECTION OF SKIN
SWEAT GLAND
HAIR FOLLICLE
ARTERY & VEIN

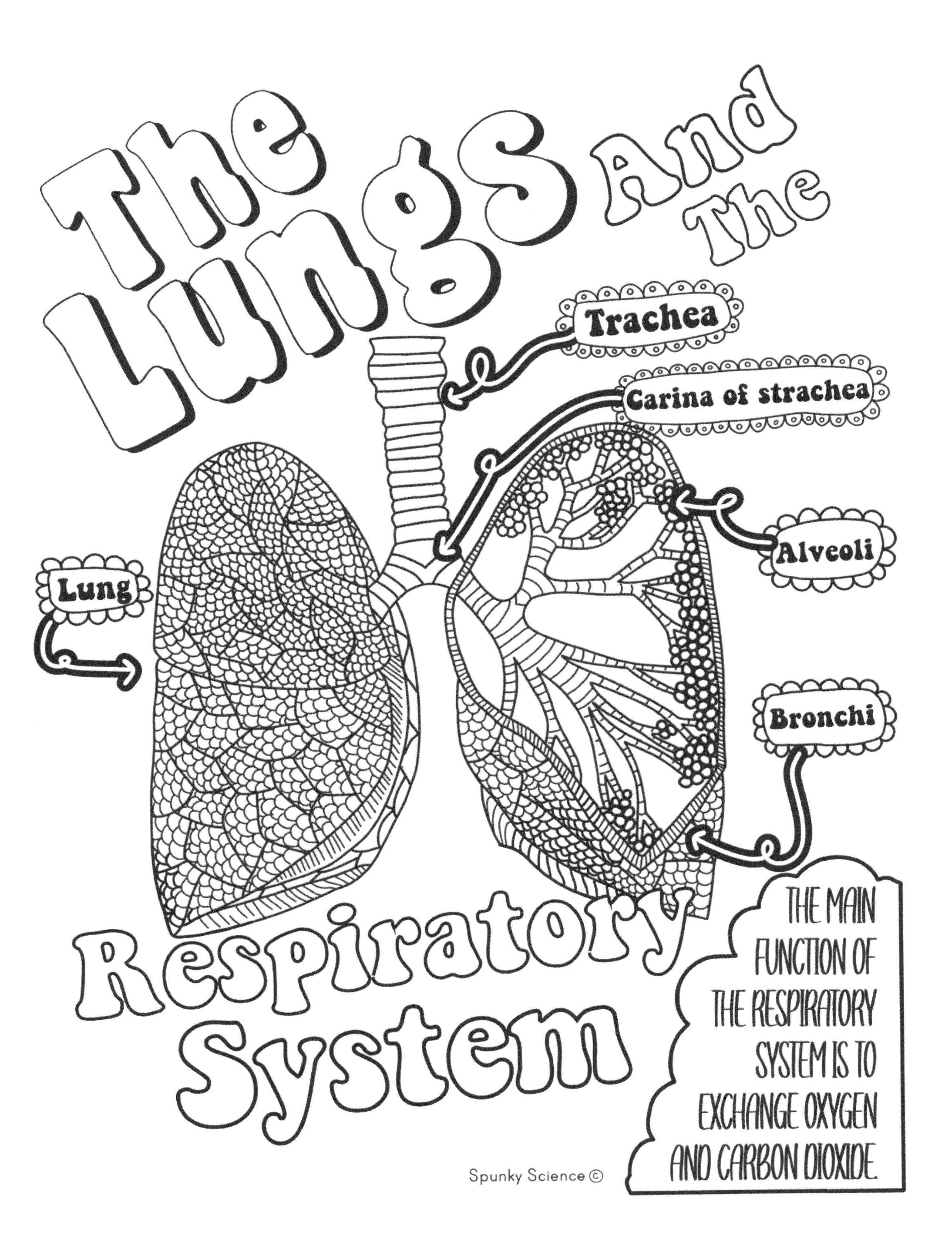
The Lungs And The
Trachea
Carina of strachea
Alveoli
Lung
Bronchi
Respiratory System
THE MAIN FUNCTION OF THE RESPIRATORY SYSTEM IS TO EXCHANGE OXYGEN AND CARBON DIOXIDE.
Spunky Science ©

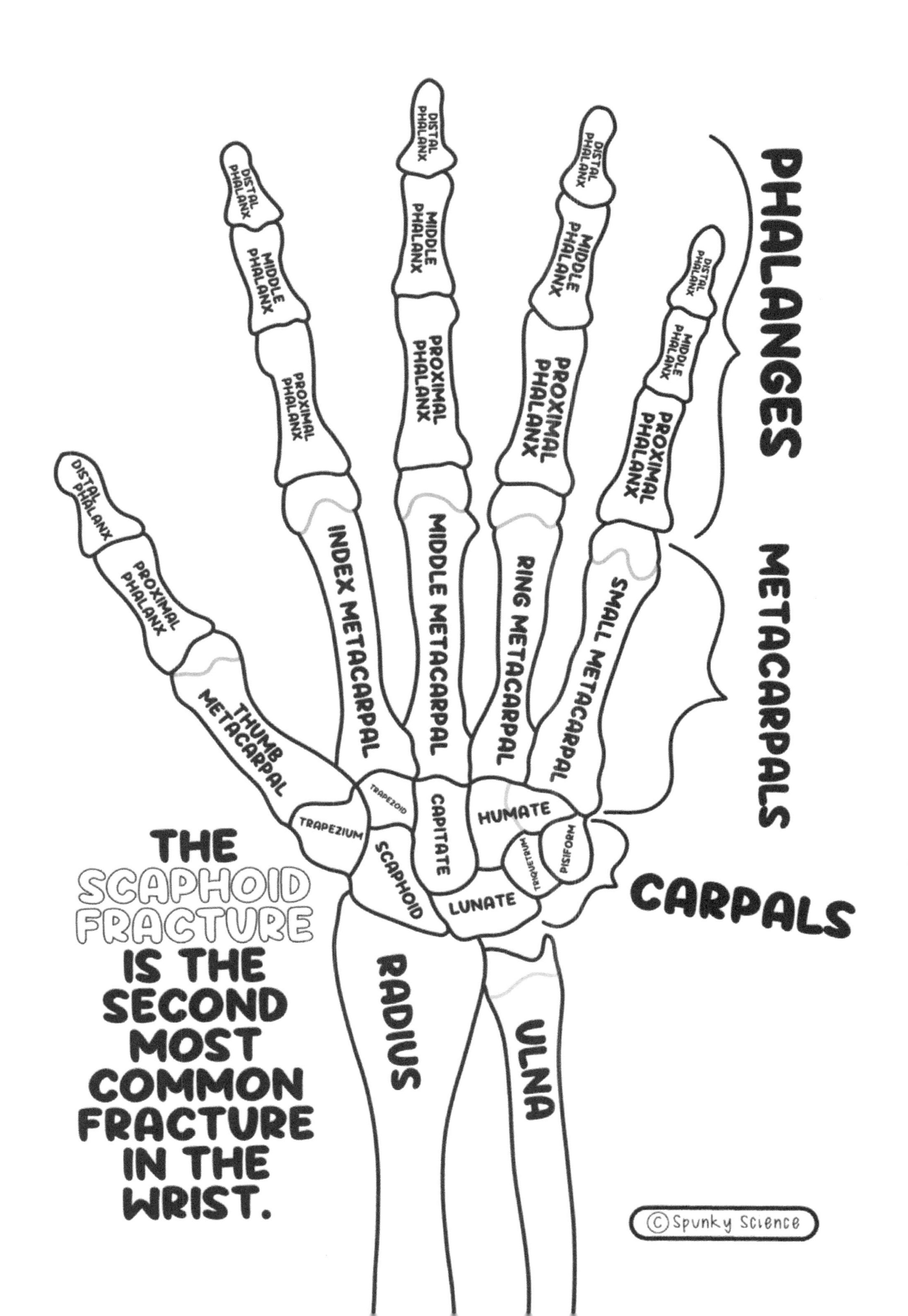
DISTAL PHALANX
MIDDLE PHALANX
PROXIMAL PHALANX
DISTAL PHALANX
MIDDLE PHALANX
PROXIMAL PHALANX
DISTAL PHALANX
MIDDLE PHALANX
PROXIMAL PHALANX
DISTAL PHALANX
MIDDLE PHALANX
PROXIMAL PHALANX
DISTAL PHALANX
PROXIMAL PHALANX
PHALANGES
THUMB METACARPAL
INDEX METACARPAL
MIDDLE METACARPAL
RING METACARPAL
SMALL METACARPAL
METACARPALS
TRAPEZOID
TRAPEZIUM
CAPITATE
HUMATE
TRIQUETRAL
PISIFORM
SCAPHOID
LUNATE
CARPALS
RADIUS
ULNA
THE
SCAPHOID
FRACTURE
IS THE
SECOND
MOST
COMMON
FRACTURE
IN THE
WRIST.

THE Skeletal SYSTEM
THE SKELETAL SYSTEM WORKS WITH MUSCLES TO PRODUCE CONTROLLED MOVEMENTS
THE SMALLEST BONE IS THE STAPES WHICH IS LOCATED IN THE MIDDLE EAR
ALSO KNOWN AS THE THIGH BONE THE FEMUR IS THE LARGEST IN THE BODY
THE MAIN FUNCTIONS OF THE SKELETAL SYSTEM ARE TO PROVIDE SUPPORT, PROTECTION, PROTECTION OF BLOOD CELLS, STORE MINERALS, AND TO REGULATE THE ENDOCRINE SYSTEM.
AT BIRTH, A BABY IS BORN WITH OVER 300 BONES! SOME OF THESE BONES FUSE TOGETHER TO FOR THE 206 BONES THAT ADULTS HAVE.
Spunky Science ©

Made in the USA
Las Vegas, NV
11 November 2023